Primarily Weather

Developed and Published
by

AIMS Education Foundation

This book contains materials developed by the AIMS Education Foundation. **AIMS** (**A**ctivities **I**ntegrating **M**athematics and **S**cience) began in 1981 with a grant from the National Science Foundation. The non-profit AIMS Education Foundation publishes hands-on instructional materials that build conceptual understanding. The foundation also sponsors a national program of professional development through which educators may gain expertise in teaching math and science.

AIMS Education Foundation
P.O. Box 8120, Fresno, CA 93747-8120 • 888.733.2467 • aimsedu.org

ISBN 978-1-60519-055-6

Printed in the United States of America

I Hear and I Forget,
I See and I Remember,
I Do and I Understand.

-Chinese Proverb

Primarily Weather

Table of Contents

Primarily Weather

All students have some kind of personal experience with weather. It influences our activities, the clothing we wear, and even where we choose to live. Just being aware of the weather builds an appreciation of our Earth's systems and the wonders of science.

Understanding the weather can better prepare our students to take precautions that will keep them and their families safe in the event of severe weather.

The activities in this book will allow students to explore many different aspects of weather, including the water cycle, temperature, wind, precipitation, storms, and the seasons and their weather patterns. To put together a comprehensive weather unit for primary students, you will find new hands-on activities, songs, and readers' theaters along with some activities that have been taken from other AIMS publications.

Learning Goals

Observing the Weather
Observe and record day-to-day weather changes
Observe and record weather conditions
Decide how to dress based on the weather

Seasons
Observe and describe seasonal changes in weather
Observe changes in plants and animals throughout the year
Relate weather to seasonal changes

Temperature
Learn how to read a thermometer
Keep temperature data throughout the year

Wind
Observe, describe, and measure the speed of the wind
Determine wind direction

Water Cycle
Identify the processes of the water cycle

Clouds
Observe and classify clouds

Precipitation
Measure precipitation
Identify different forms of precipitation

Severe Weather
Identify severe weather events
Learn safety precautions to be used in severe weather

National Education Reform Documents

The AIMS Education Foundation is committed to remaining at the cutting edge of providing curricular materials that are user-friendly, educationally sound, developmentally appropriate, and aligned with the recommendations found in national education reform documents.

Project 2061 Benchmarks

The Nature of Science

- *People can often learn about things around them by just observing those things carefully, but sometimes they can learn more by doing something to the things and noting what happens.*
- *Tools such as thermometers, magnifiers, rulers, or balances often give more information about things than can be obtained by just observing things without their help.*
- *Describing things as accurately as possible is important in science because it enables people to compare their observations with those of others.*

The Nature of Mathematics

- *Things move, or can be made to move, along straight, curved, circular, back-and-forth, and jagged paths.*

The Nature of Technology

- *Tools are used to do things better or more easily and to do some things that could not otherwise be done at all. In technology, tools are used to observe, measure, and make things.*

The Physical Setting

- *The sun can be seen only in the daytime, but the moon can be seen sometimes at night and sometimes during the day. The sun, moon, and stars all appear to move slowly across the sky.*
- *The moon looks a little different every day, but looks the same again about every four weeks.*
- *Some events in nature have a repeating pattern. The weather changes some from day to day, but things such as temperature and rain (or snow) tend to be high, low, or medium in the same months every year.*
- *Water can be a liquid or a solid and can go back and forth from one form to the other. If water is turned into ice and then the ice is allowed to melt, the amount of water is the same as it was before freezing.*
- *Water left in an open container disappears, but water in a closed container does not disappear.*
- *Change is something that happens to many things.*
- *The sun warms the land, air, and water.*
- *Things move in many different ways, such as straight, zigzag, round and round, back and forth, and fast and slow.*
- *When liquid water disappears, it turns into a gas (vapor) in the air and can reappear as a liquid when cooled, or as a solid if cooled below the freezing point of water. Clouds and fog are made of tiny droplets of water.*

The Living Environment

- *Plants and animals have features that help them live in different environments.*

The Human Organism

- *People use their senses to find out about their surroundings and themselves. Different senses give different information. Sometimes a person can get different information about the same thing by moving closer to it or further away from it.*

The Mathematical World

- *Simple graphs can help to tell about observations.*
- *Some things are more likely to happen than others. Some events can be predicted well and some cannot. Sometimes people aren't sure what will happen because they don't know everything that might be having an effect.*

Common Themes

- *A model of something is different from the real thing but can be used to learn something about the real thing.*
- *Things change in some ways and stay the same in some ways.*

Habits of Mind

- *Raise questions about the world round them and be willing to seek answers to some of them by making careful observations and trying things out.*
- *Draw pictures that correctly portray at least some features of the thing being described.*

American Association for the Advancement of Science
Benchmarks for Science Literacy
Oxford University Press. New York. 1993.

NRC Standards

Abilities Necessary to do Scientific Inquiry
- *Ask a question about objects, organisms, and events in the environment.*
- *Employ simple equipment and tools to gather data and extend the senses.*
- *Communicate investigations and explanations.*

Understandings About Scientific Inquiry
- *Simple instruments, such as magnifiers, thermometers, and rulers, provide more information than scientists obtain using only their senses.*

Properties of Objects and Materials
- *Objects have many observable properties, including size, weight, shape, color, temperature, and the ability to react with other substances. Those properties can be measured using tools, such as rulers, balances, and thermometers.*
- *Materials can exist in different states—solid, liquid, and gas. Some common materials, such as water, can be changed from one state to another by heating or cooling.*

The Characteristics of Organisms
- *The behavior of individual organisms is influenced by internal cues (such as hunger) and by external cues (such as a change in the environment). Humans and other organisms have senses that help them detect internal and external cues.*

Objects in the Sky
- *The sun, moon, stars, clouds, birds, and airplanes all have properties, locations, and movements that can be observed and described.*
- *The sun provides the light and heat necessary to maintain the temperature of the earth.*

Changes in the Earth and Sky
- *Weather changes from day to day and over the seasons. Weather can be described by measurable quantities, such as temperature, wind direction and speed, and precipitation.*
- *Objects in the sky have patterns of movement. The sun, for example, appears to move across the sky in the same way every day, but its path changes slowly over the seasons. The moon moves across the sky on a daily basis much like the sun. The observable shape of the moon changes from day to day in a cycle that lasts about a month.*

Structure of the Earth System
- *Water, which covers the majority of the earth's surface, circulates through the crust, oceans, and atmosphere in what is known as the "water cycle." Water evaporates from the earth's surface, rises and cools as it moves to higher elevations, condenses as rain or snow, and falls to the surface where it collects in lakes, oceans, soil, and in rocks underground.*

Understanding About Science and Technology
- *Tools help scientists make better observations, measurements, and equipment for investigations. They help scientists see, measure, and do things that they could not otherwise see, measure, and do.*

Personal Health
- *Safety and security are basic needs of humans. Safety involves freedom from danger, risk, or injury. Security involves feelings of confidence and lack of anxiety and fear. Student understandings include following safety rules for home and school, preventing abuse and neglect, avoiding injury, knowing whom to ask for help, and when and how to say no.*

National Research Council
National Science Education Standards
National Academy Press. Washington, DC. 1996.

Common Core Standards for Math

Kindergarten
Counting and Cardinality
- *Know number names and the count sequence.*
- *Count to tell the number of objects.*

Measurement and Data
- *Classify objects and count the number of objects in each category.*

First Grade
Measurement and Data
- *Measure lengths indirectly and by iterating length units.*
- *Represent and interpret data.*

Second Grade
Measurement and Data
- *Measure and estimate lengths in standard units.*
- *Work with time and money.*
- *Represent and interpret data.*

Third Grade
Measurement and Data
- *Represent and interpret data.*

Calendar Connections

Learning Goals
Students will:
- observe and record day-to-day weather changes (hot, cold, raining, etc.);
- observe and describe seasonal changes in weather by selecting an appropriate weather wear picture;
- observe and record whether the sun and moon can be seen during the day;
- identify and extend patterns on the calendar,
- recognize numbers; and
- experience one-to one correspondence.

Guiding Documents
Project 2061 Benchmarks
- *Tools such as thermometers, magnifiers, rulers, or balances often give more information about things than can be obtained by just observing things without their help.*
- *Describing things as accurately as possible is important in science because it enables people to compare their observations with those of others.*
- *The sun can be seen only in the daytime, but the moon can be seen sometimes at night and sometimes during the day. The sun, moon, and stars all appear to move slowly across the sky.*
- *Some events in nature have a repeating pattern. The weather changes some from day to day, but things such as temperature and rain (or snow) tend to be high, low, or medium in the same months every year.*
- *The moon looks a little different every day, but looks the same again about every four weeks.*

NRC Standards
- *Simple instruments, such as magnifiers, thermometers, and rulers, provide more information than scientists obtain using only their senses.*
- *The sun, moon, stars, clouds, birds, and airplanes all have properties, locations, and movements that can be observed and described.*
- *The sun provides the light and heat necessary to maintain the temperature of the earth.*
- *Weather changes from day to day and over the seasons. Weather can be described by measurable quantities, such as temperature, wind direction and speed, and precipitation.*
- *Objects in the sky have patterns of movement. The sun, for example, appears to move across the sky in the same way every day, but its path changes slowly over the seasons. The moon moves across the sky on a daily basis much like the sun. The observable shape of the moon changes from day to day in a cycle that lasts about a month.*
- *Tools help scientists make better observations, measurements, and equipment for investigations. They help scientists see, measure, and do things that they could not otherwise see, measure, and do.*

*Common Core Standards for Math**
- *Know number names and the count sequence. (K.CC)*
- *Count to tell the number of objects. (K.CC)*
- *Classify objects and count the number of objects in each category. (K.MD)*
- *Represent and interpret data. (1.MD/2.MD/3.MD)*

Materials
Calendar cut outs
Large paper thermometer (see *Management 1)*
Weather wear pictures (see *Management 2)*
Teeth
Birthday cake and candles
Days of the week on sentence strips
Large sun
Large moon

Background Information

Calendar time is the time of day in a primary classroom that many math and science skills can be introduced and even mastered through daily repetition. Counting forward, backward, skip-counting, and weather awareness can be learned very quickly when repeated each day in a routine that the students can enjoy. Suggested items to include on the calendars are as follows:

The Calendar

The focus of this section in the beginning of the year should be on number recognition, naming the days of the week in order, and identifying simple patterns. As the year progresses, more complex patterns can be introduced as well as skip counting by twos, fives, and tens.

Monthly Birthdays and Teeth Lost

In this section of the calendar, recognize those students with a birthday during the month by placing a candle with the student's name onto the cake. At the end of each month, transfer the cake and candles to the wall where the class can place the months in order and create a graph to analyze. Keep track of the number of lost teeth each month by writing the name of each child who loses a tooth during the month on the large tooth. This monthly information can also be transferred to create a pictograph that compares the number of teeth lost each month.

Time Vocabulary—Today, Yesterday, Tomorrow

Introduce the term *yesterday* by reviewing what day of the week yesterday was and what special activities or events occurred. For example, yesterday some students may have gotten ready for the first day of school by going to the store to get school supplies or by going to bed early. Introduce the term *today* by talking about what day of the week it is. Discuss that today is the first day of school and explain any special activities that will occur, especially those that will become part of the regular school schedule, such as art on Thursdays. Introduce the term *tomorrow* by talking about what day of the week tomorrow will be. Announce any special activities that will take place, such as a regularly scheduled activity (e.g., music). Also explain that tomorrow will be the second day of school. Select a student to place the correct day of the week into the appropriate sentence.

Seasonal Changes

This section of the calendar provides an opportunity for children to see that some events in nature have a repeating pattern. The weather changes from day to day, but things such as temperature tend to be *high, low,* or *medium* in the same months every year. After students have a basic understanding about the appearance of a thermometer, introduce the class to the word *meteorologist.* Explain that a meteorologist studies weather patterns and predicts weather. Tell the class that the job of classroom meteorologist will be added to the weekly job list. Explain that during calendar time, the class meteorologist will use the wall thermometer to show if the weather outside is hot, cold, or somewhere in between each day. It should be part of his or her job to report the weather to the class by saying it is _____ today, yesterday it was _____, and tomorrow I think it will be _____. It will also be the responsibility of the meteorologist to decide and post the appropriate weather wear for the conditions of the day.

Sun and Moon

This section of the calendar provides an opportunity for the students to discover that the sun can be seen only in the daytime, but the moon can be seen sometimes at night and sometimes during the day. Each day during calendar time it will be one child's job to be the sky watcher. He or she will need to step to the window or outside to see if the sun, moon, or both are visible. He or she will then report the information to the class and record a tally mark under the appropriate heading(s). You may choose to vary the time that this observation is made based when the moon will be visible that day. (This information can be found on the Internet at the Weather Underground: www.wunderground.com.)

Management

1. Copy the thermometer pieces onto railroad board and tape them together. Laminate for extended use. Cut the appropriate slots in the thermometer. Cut one 2" x 20" strip of red paper or ribbon and one 2" x 12" white strip that is similar in weight. Tape the strips together to form one continuous strip and laminate. Place the strip through the slots in the railroad board thermometer and tape them together, forming a loop on the underside of the thermometer so that they slide freely to represent highs and lows in temperature.
2. Prior to implementing the weather wear section of the calendar, take photos of the students wearing the appropriate clothing for various types of weather such as rainy, cold, hot, etc. Laminate for extended use.

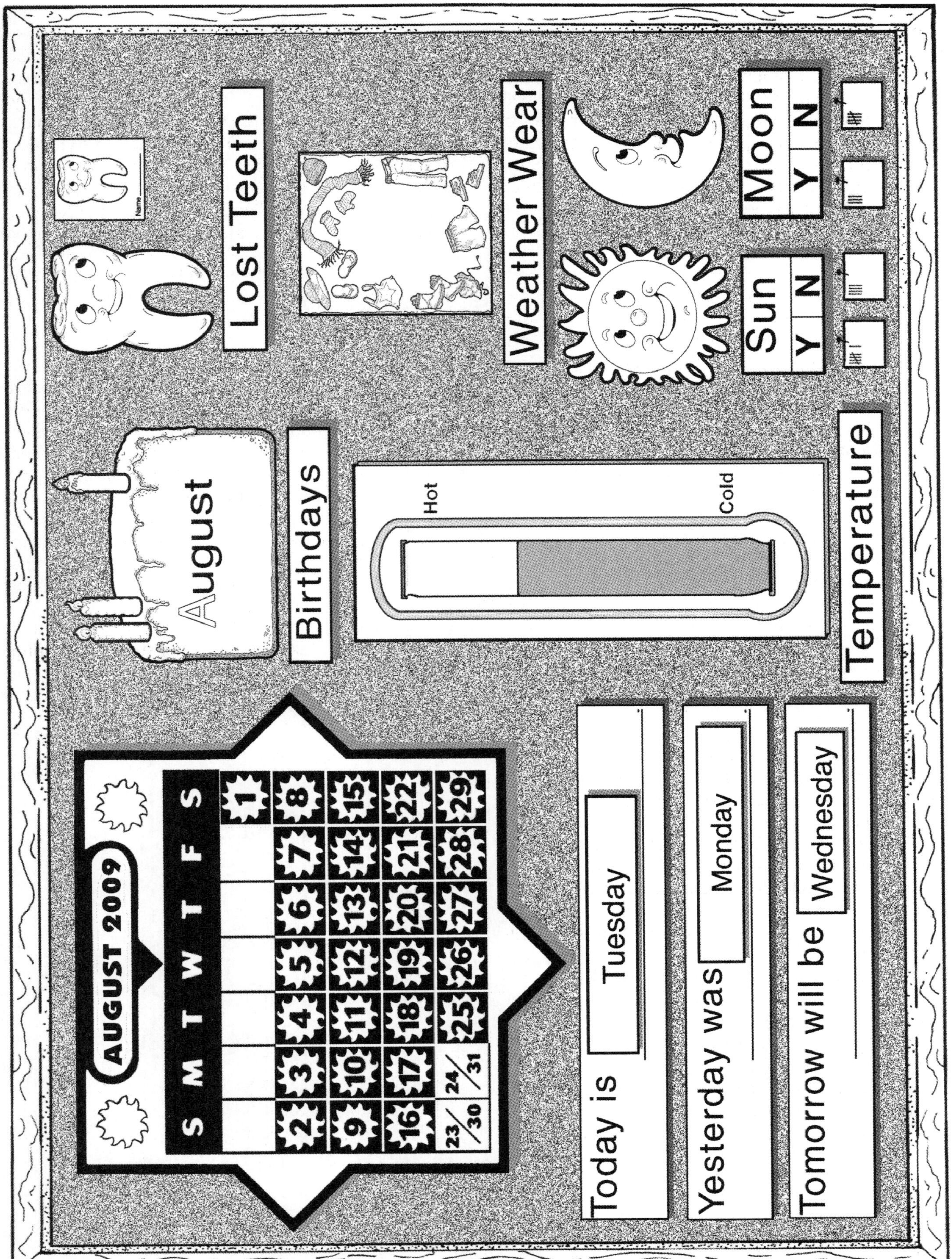
AUGUST 2009
S M T W T F S
1 2 3 4 5 6 7 8 9 10 11 12 13 14 15 16 17 18 19 20 21 22 23/30 24/31 25 26 27 28 29
Today is Tuesday
Yesterday was Monday
Tomorrow will be Wednesday
August
Birthdays
Hot
Cold
Temperature
Lost Teeth
Name
Weather Wear
Sun
Y N
Moon
Y N

Thermometer Assembly

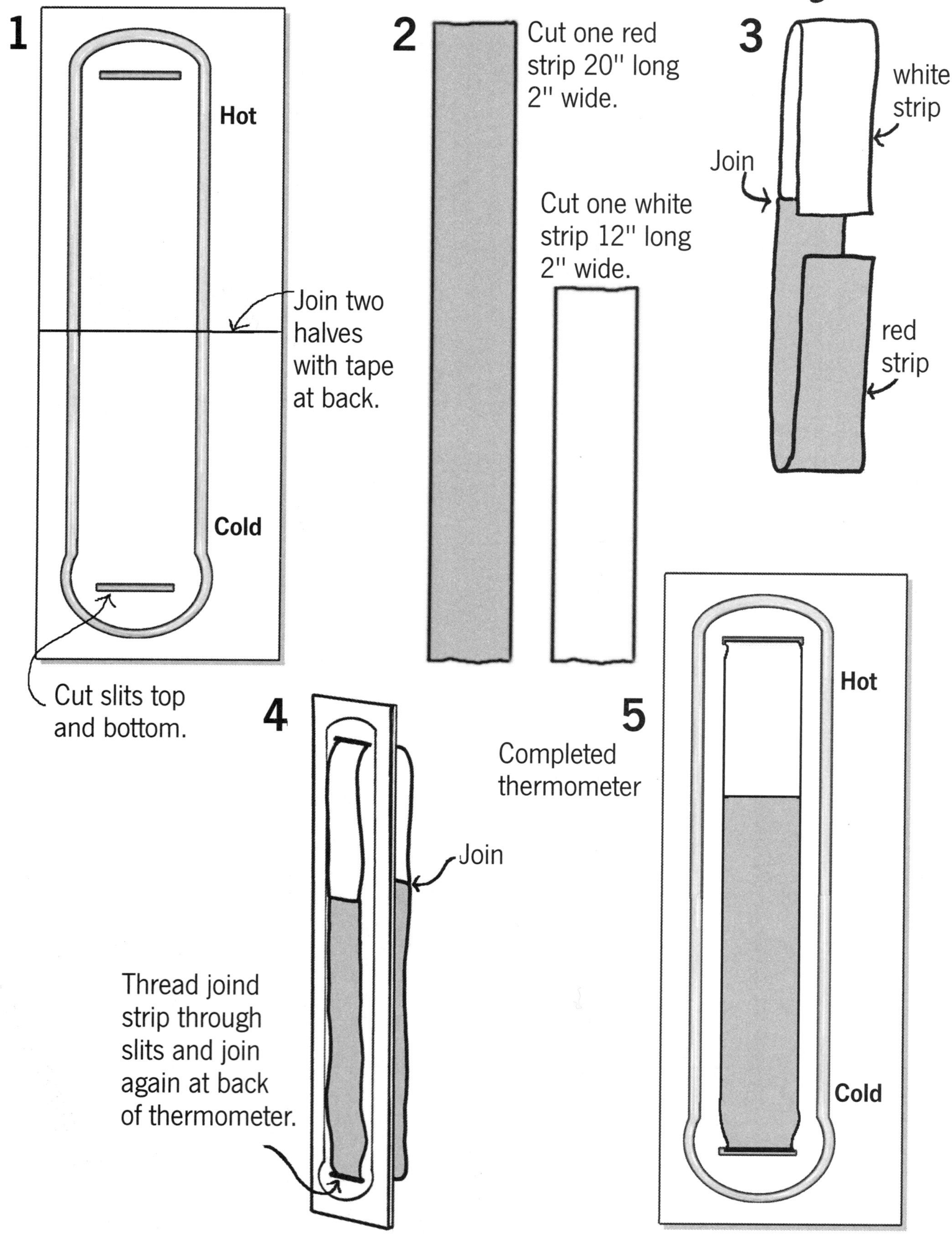

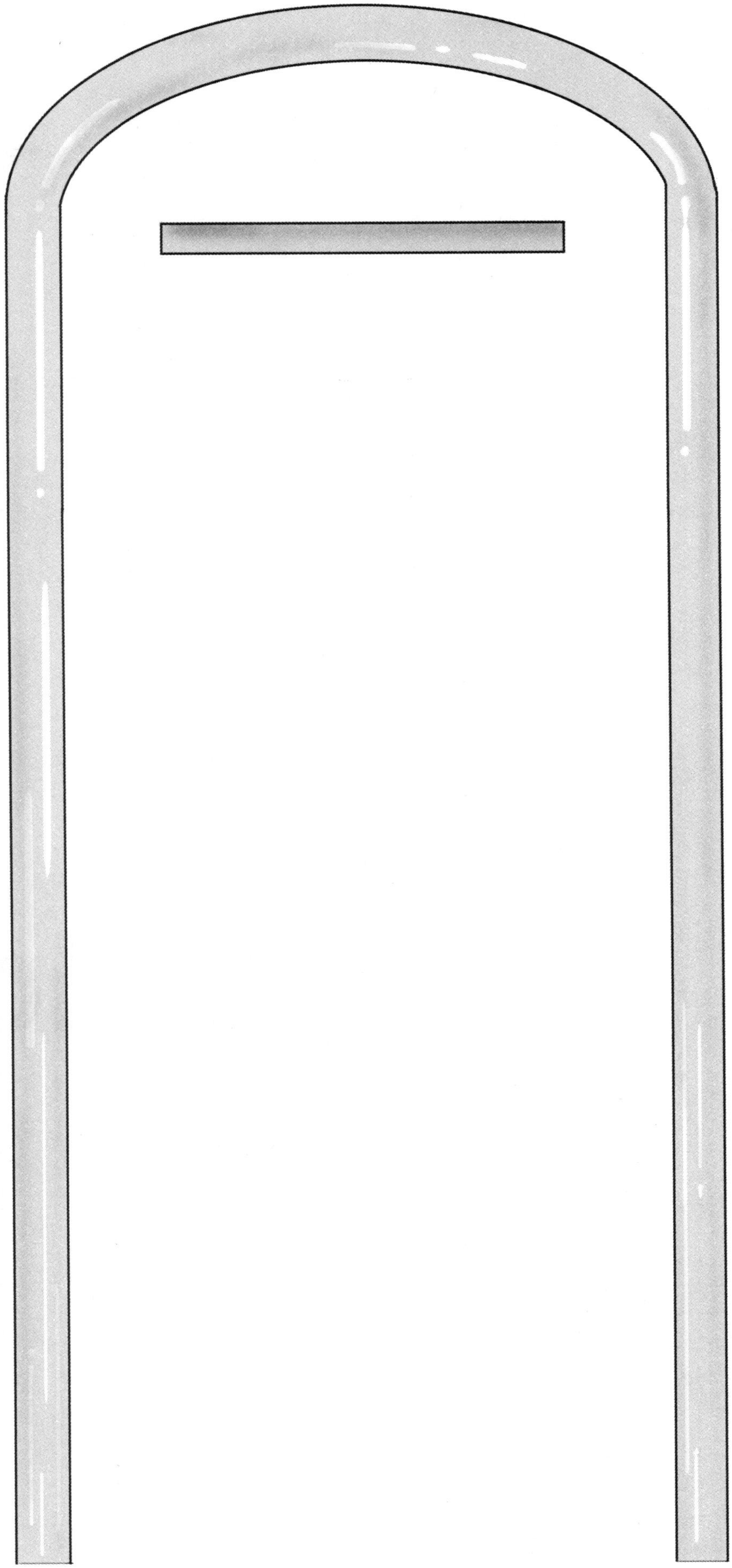

Hot

Cold

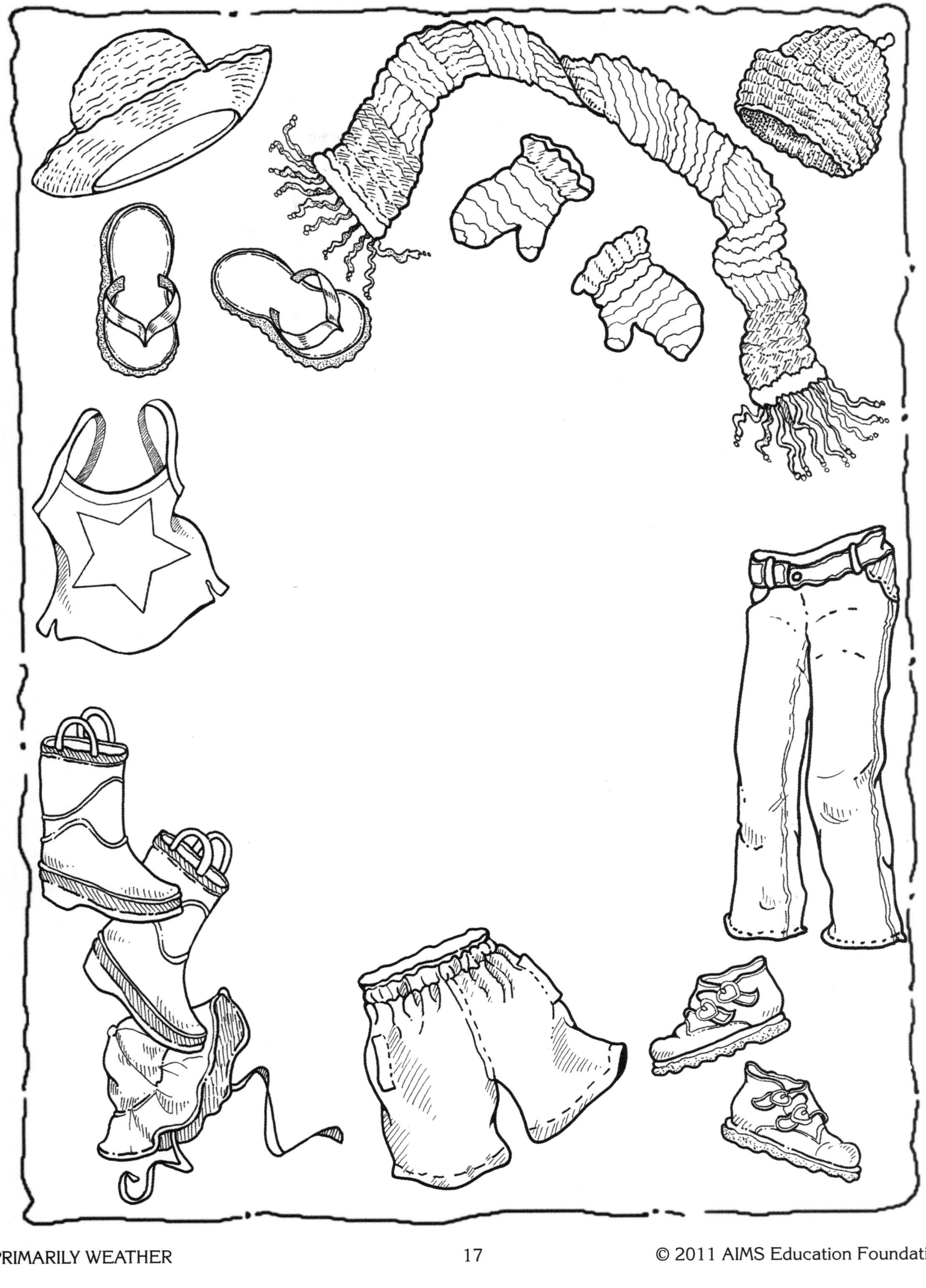

Moon	
Y	N

Sun	
Y	N

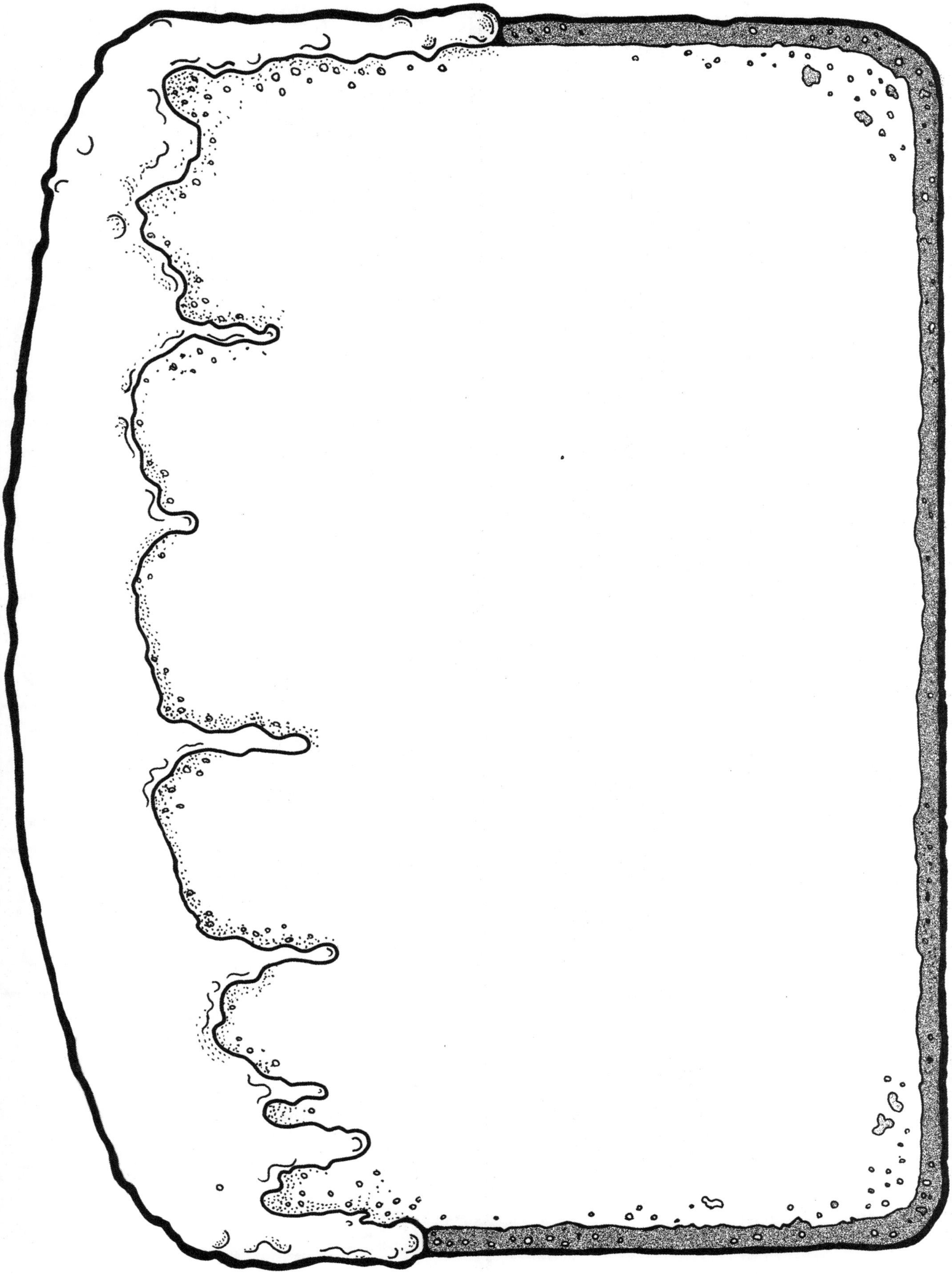

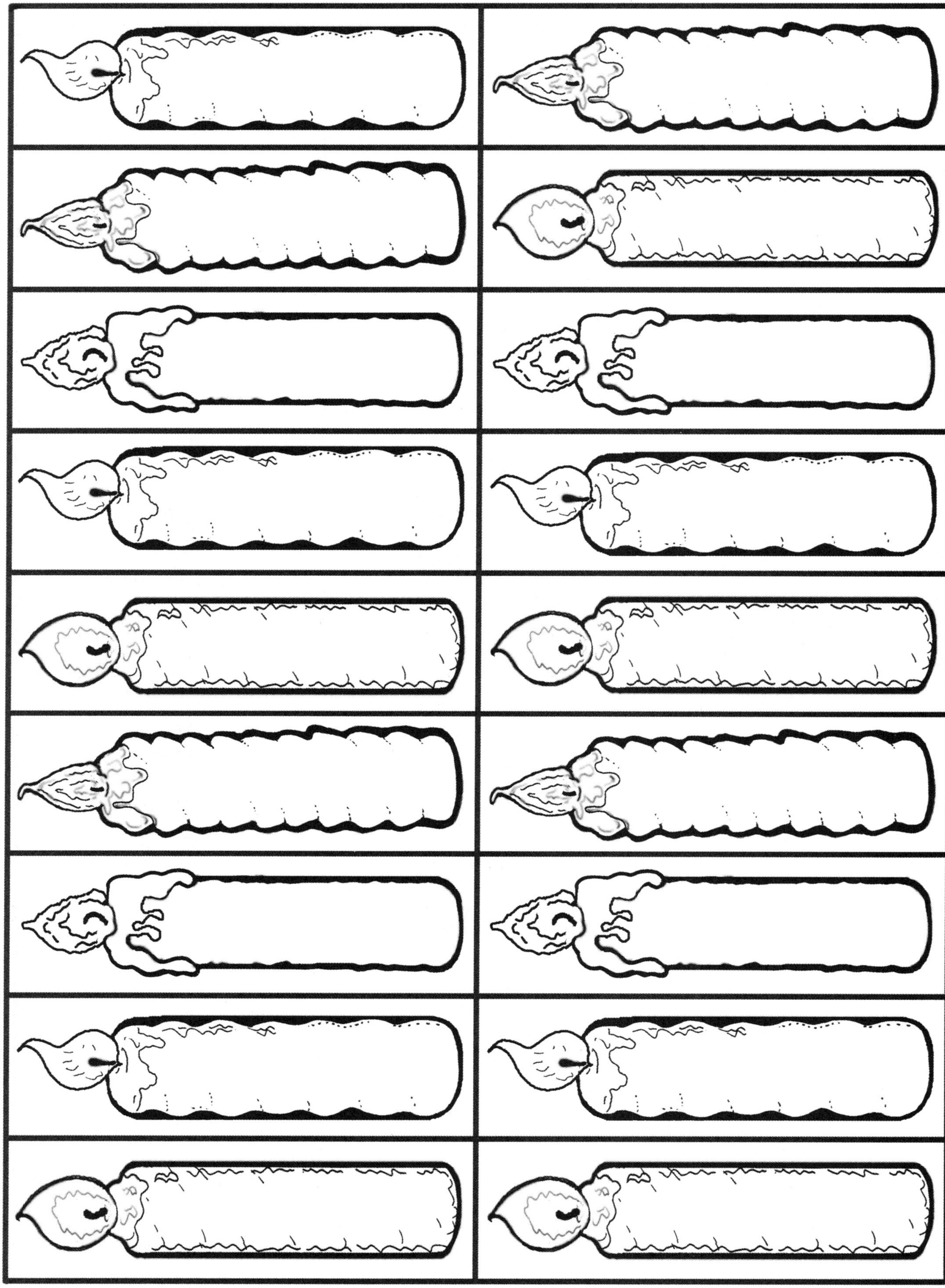

January	February	March

April

May

June

July	August	September

October	November	December

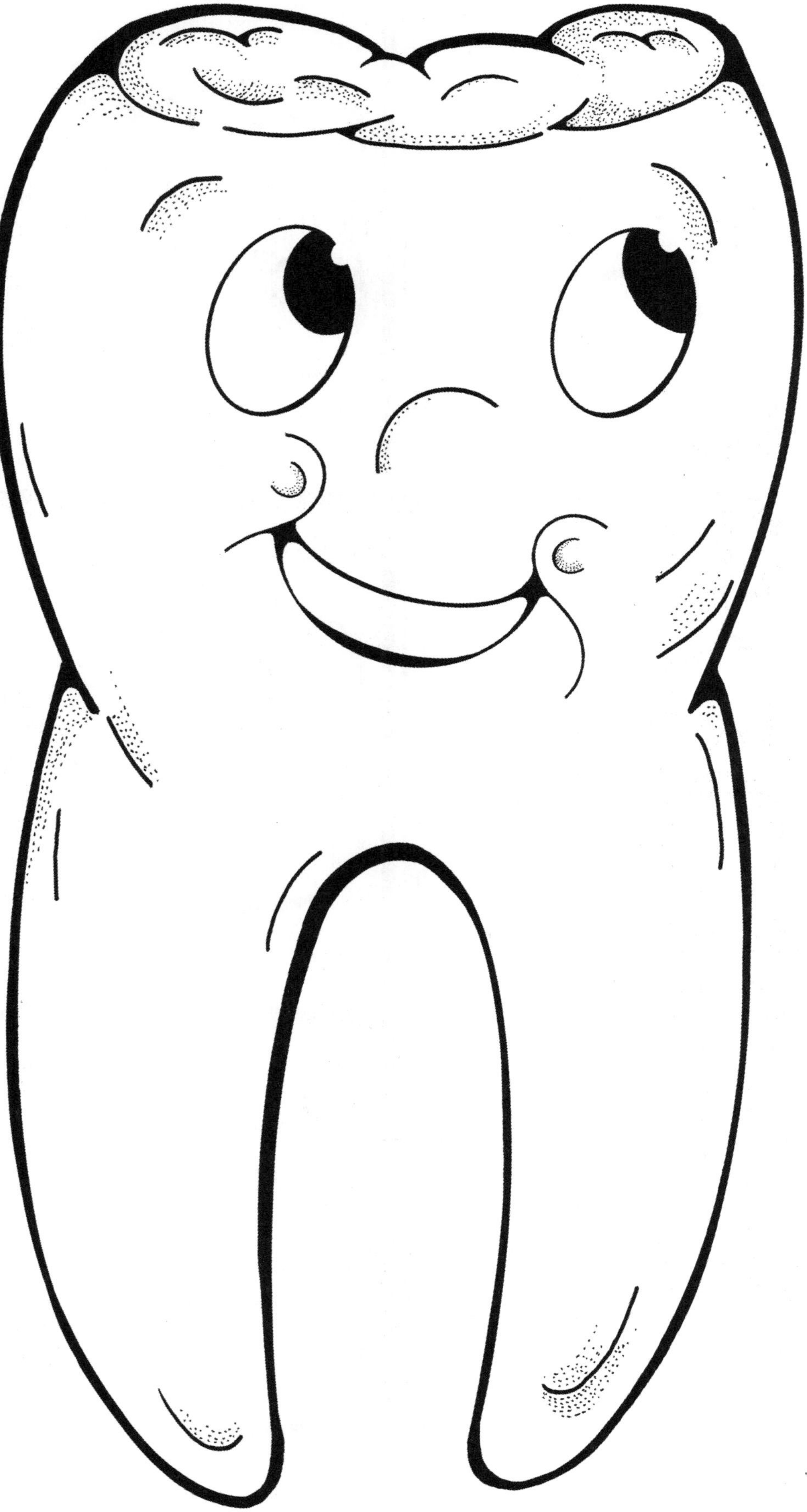

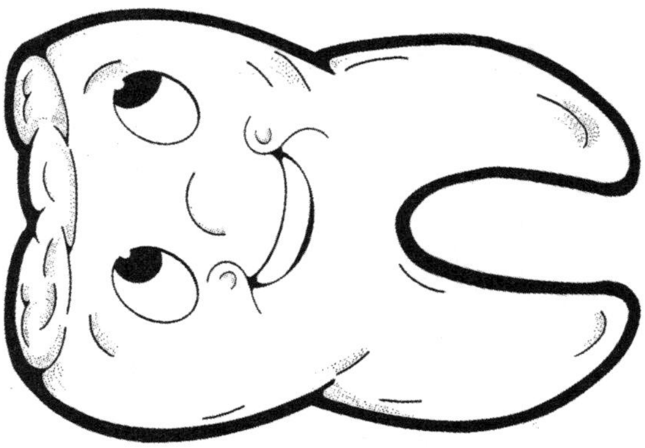

Name ______

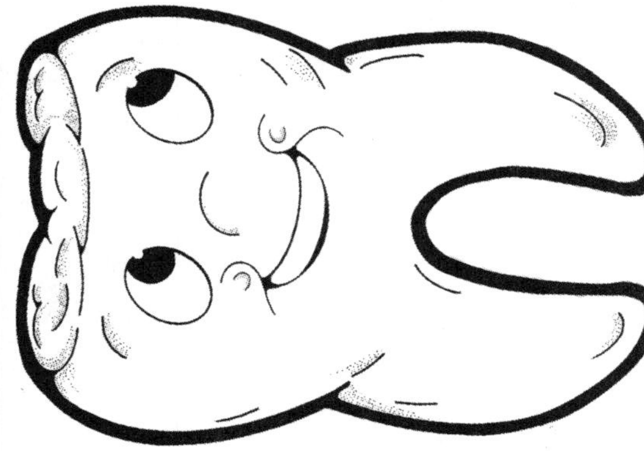

Name ______

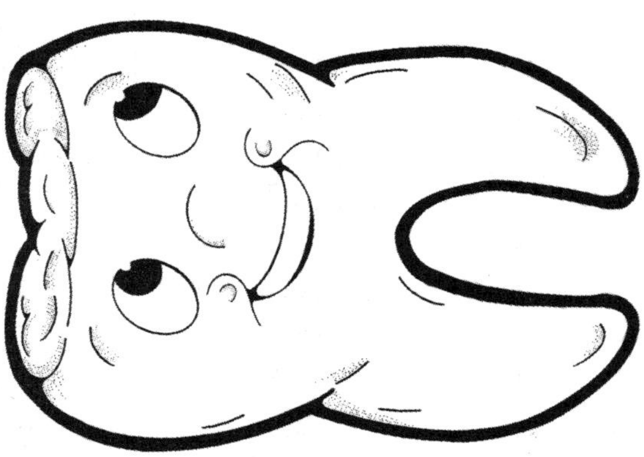

Name ______

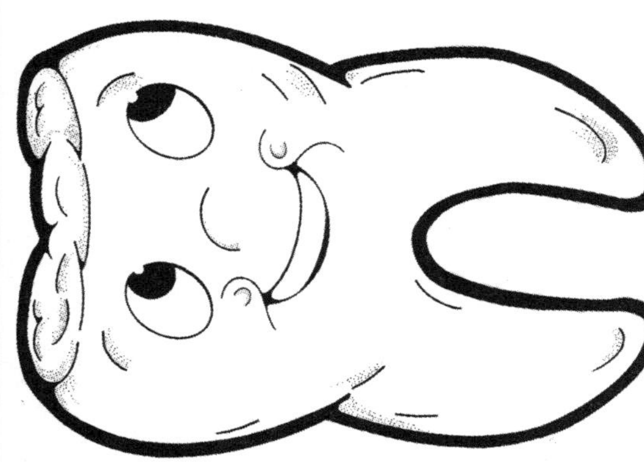

Name ______

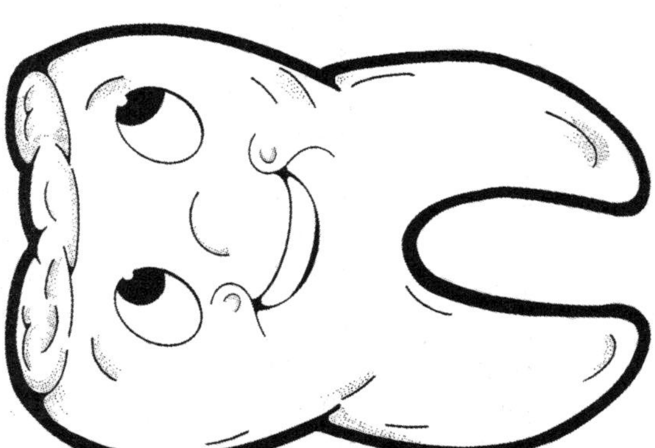

Name ______

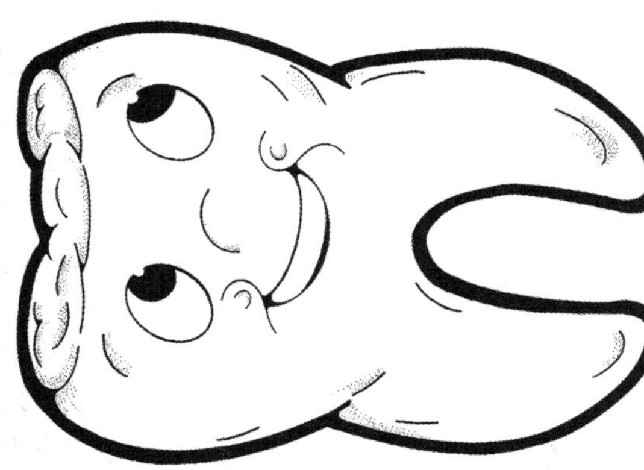

Name ______

Today is ______________________.

Yesterday was ______________________.

Tomorrow will be ______________________.

Monday
Tuesday
Wednesday
Thursday
Friday
Saturday
Sunday

A Tool for Temperature

Topic
Reading a thermometer

Key Question
How is a thermometer like a number line?

Learning Goals
Students will:
- learn how to read a thermometer,
- use a thermometer to gather scientific data, and
- measure properties of objects using thermometers.

Guiding Documents
Project 2061 Benchmark
- *Tools such as thermometers, magnifiers, rulers, or balances often give more information about things than can be obtained by just observing things without their help.*

NRC Standards
- *Simple instruments, such as magnifiers, thermometers, and rulers, provide more information than scientists obtain using only their senses.*
- *Employ simple equipment and tools to gather data and extend the senses.*

Math
Measurement
  temperature

Science
Earth science
  weather
    temperature
Physical science
  heat energy

Integrated Processes
Observing
Collecting and recording data
Comparing and contrasting
Interpreting data

Materials
Class number lines (see *Management 1)*
Thermometers
Student pages
Sticky notes (see *Management 2)*
Class thermometer (see *Management 3)*

Background Information
It is important for students to understand that a thermometer is an instrument used for measuring temperature. Since heat is measured in degrees, a thermometer tells us how many degrees of heat something has. It will be necessary for students to recognize number patterns and fill in missing numbers as they identify exact temperatures since various thermometers use different scales. To assist students in reading and understanding various scales, *Part One* of this activity reviews horizontal number line skills and transitions to a vertical number line similar to what is seen on a thermometer.

In *Part Two* of the activity, students will learn how to use a thermometer to determine temperatures that will become part of the daily calendar procedures.

Management
1. Copy, assemble, and display the number lines included in this activity. Laminate for extended use.
2. When focusing on five- and 10-point scales, cover the other numbers with sticky notes.
3. Copy the thermometer pieces and glue onto railroad board. Laminate for extended use. Construct the thermometer using the *Thermometer Assembly* directions.
4. Divide the class into groups of three for *Part Two* of this activity.
5. Gather one thermometer per group of three students. Thermometers (item number 1976) are available from AIMS.
6. Prior to copying, write the scale on the student page of thermometers. Alternately, students can record the scale as well as the correct temperatures.

Procedure
Part One—Vertical number line
1. Direct the students' attention to the horizontal number line. Discuss what the counting pattern is. [increasing by one] Using sticky notes, cover several numbers and have the students identify what numbers are missing.
2. Cover all numbers except the multiples of five [5, 10, 15, etc.] and ask the students to identify the scale. Point to several covered numbers and ask the students to identify the missing numbers. Repeat the process covering all numbers except the multiples of 10.

3. When students have demonstrated a good understanding of the horizontal number line, introduce the vertical number line. Ask the students how the two number lines are similar and how they are different.
4. Repeat the covering and questioning process used in *Procedures 1* and *2.*
5. Discuss situations where the students might have to use a horizontal and/or a vertical number line in the real world.

Part Two—Using a thermometer

1. Display the class thermometer. Explain that this will be a model thermometer. Ask the students to make observations. If the students do not mention the similarity between the thermometer and the number lines they previously worked with, draw their attention to the numbers up the side, and ask them how the numbers going up the side of the thermometer are similar to the ones on their number line. Tell students that these numbers are called the scale and that temperature is measured in *degrees.*
2. Point to several positions that fall between numbers and have the students identify the missing number on the model thermometer.
3. Draw students' attention to the red paper. Explain that most real thermometers contain a liquid (mercury or alcohol) that expands when heated. As it expands, it rises up the tube. As the liquid inside the thermometer cools, it falls. Discuss what the thermometer might look like on a very hot day. Invite a student to adjust the red paper in the class thermometer so that it represents a high temperature.
4. Distribute one thermometer to each group. Ask students to compare this thermometer with the class thermometer.
5. Show the students that the top of the liquid indicates the temperature.
6. Have the students read the numbers actually appearing on the scale. Discuss how many degrees each line on the thermometer represents.
7. Ask the students to identify the highest and lowest temperatures they could have on their thermometers.
8. Have the students find the temperature inside the classroom. Ask each group to record the temperature on the student page. They will need to determine an appropriate scale for the thermometer and write in the numbers along the side. Be sure they indicate whether the measurements are in Celsius or Fahrenheit. (The class thermometer shows a Fahrenheit scale because this is the scale students will hear and see through various media reports.)
9. Take the class outside. Instruct each group to place its thermometer in a shady area for three minutes, take a reading, and record it on the student page. Have students predict what the temperature might be before placing their thermometers.
10. Have the students place their thermometers in a sunny spot, predict what the temperature will be, and record the temperature after three minutes. Discuss the results. Encourage students to give reasons for differences in temperatures among the groups.

Connecting Learning

1. How is a thermometer scale a number line?
2. How do you read a thermometer?
3. What does a thermometer measure?
4. What might the thermometer look like on a very cold day? ...a very hot day?
5. Do the thermometers of all of the groups show the same temperature? Explain your thinking.
6. What else would you like to know about a thermometer?
7. Of the three temperatures you recorded, which was the hottest? How do you know?
8. What are you wondering now?

Thermometer Assembly

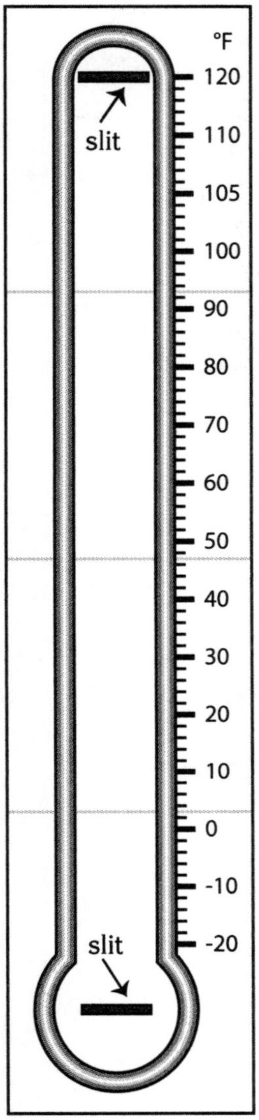

1. Cut along the solid lines and join the four pages as indicated. Cut slits in the top and bottom of the thermometer where indicated by the rectangles.

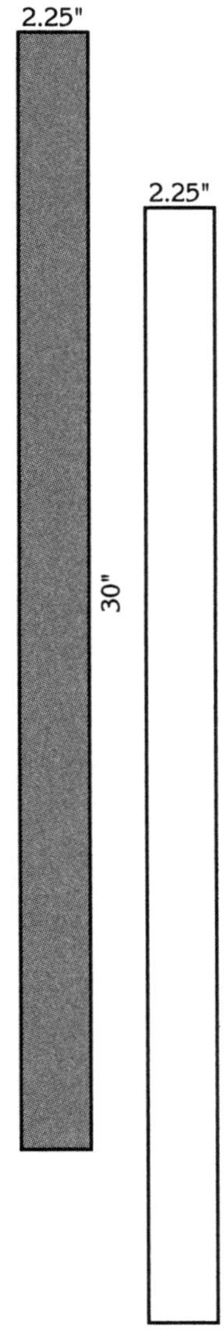

2. Cut a strip of adding machine tape 30" long. Cut a matching strip of red butcher paper 2 ¼" wide and 30" long.

3. Tape the ends of the red and white strips together to make one long strip. The ends can overlap a bit.

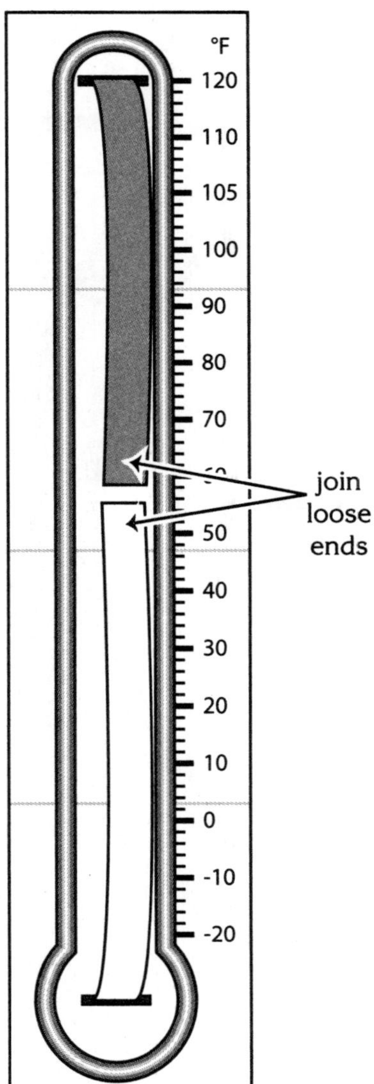

4. Thread the joined strip through the slits in the thermometer and join the loose ends with tape.

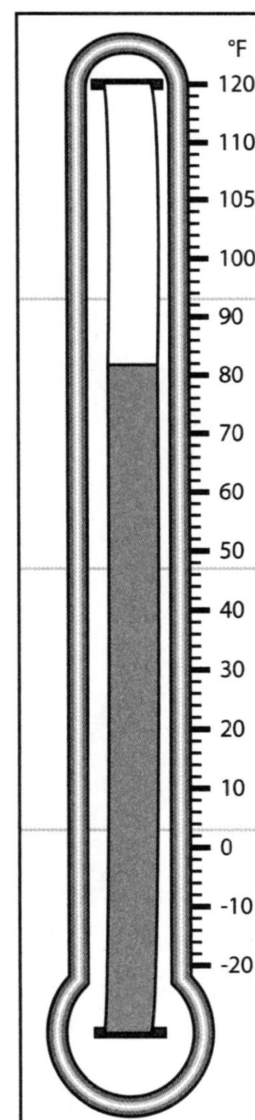

5. This is the completed thermometer.

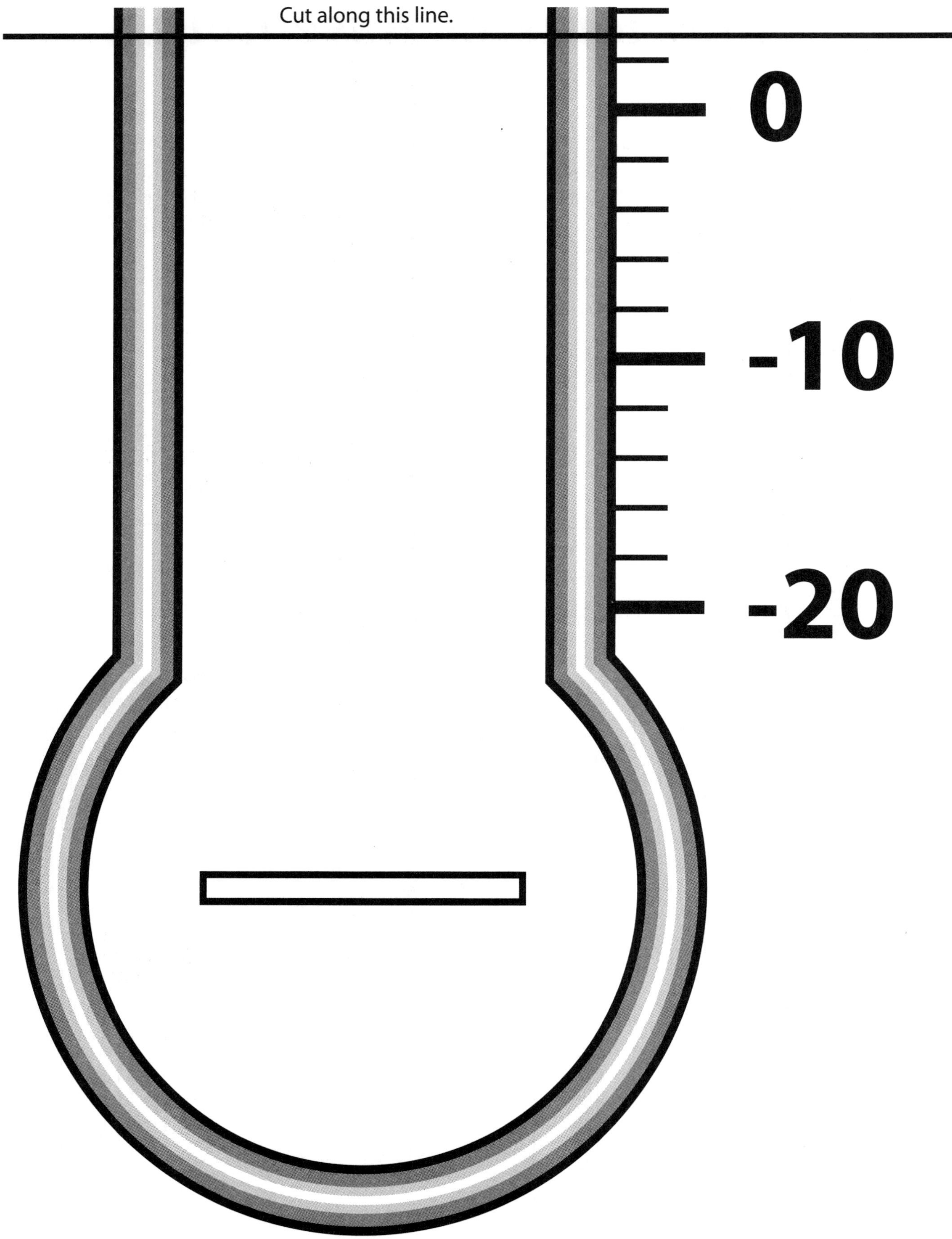
Cut along this line.
0
-10
-20

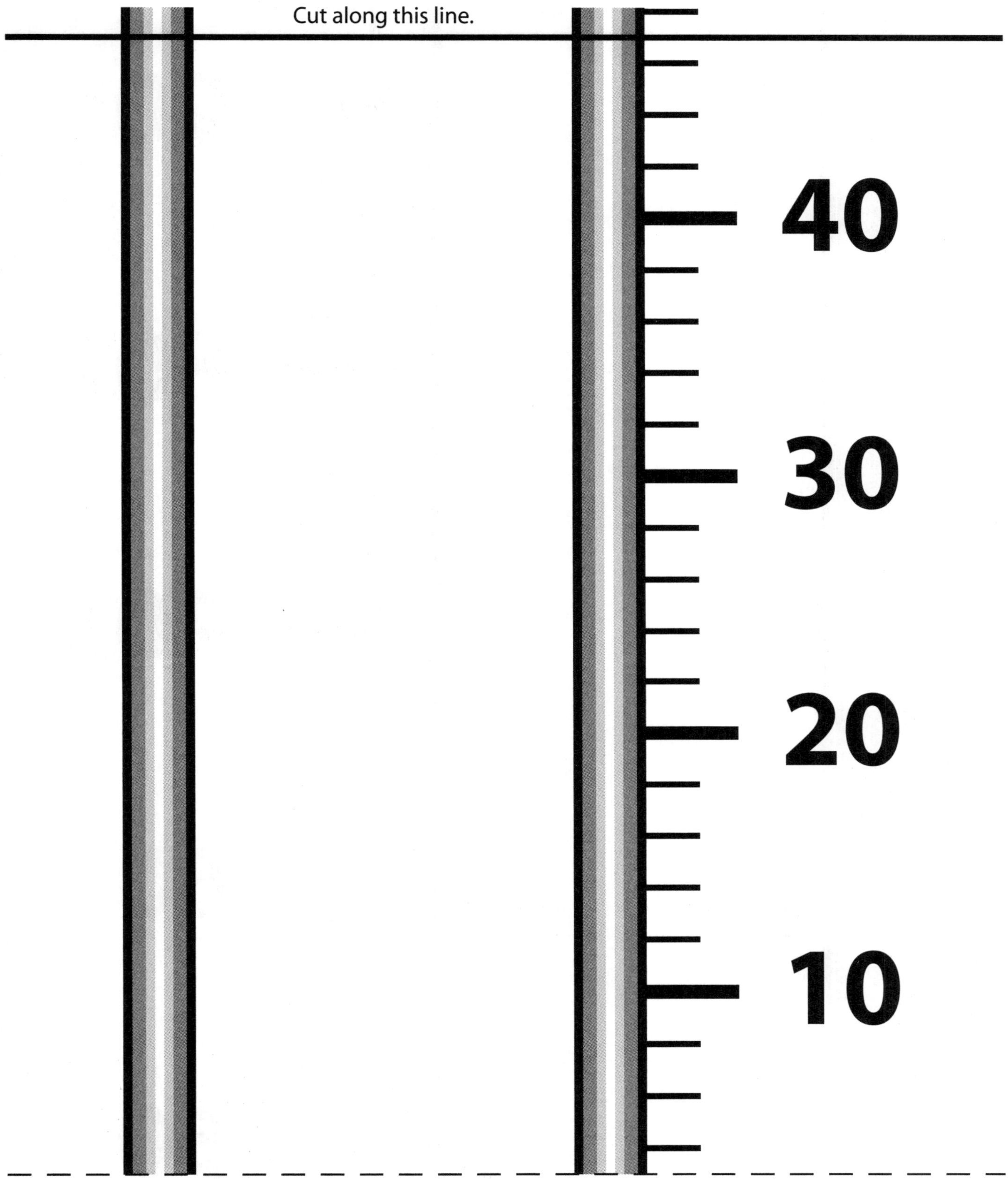

Tape the first page here.

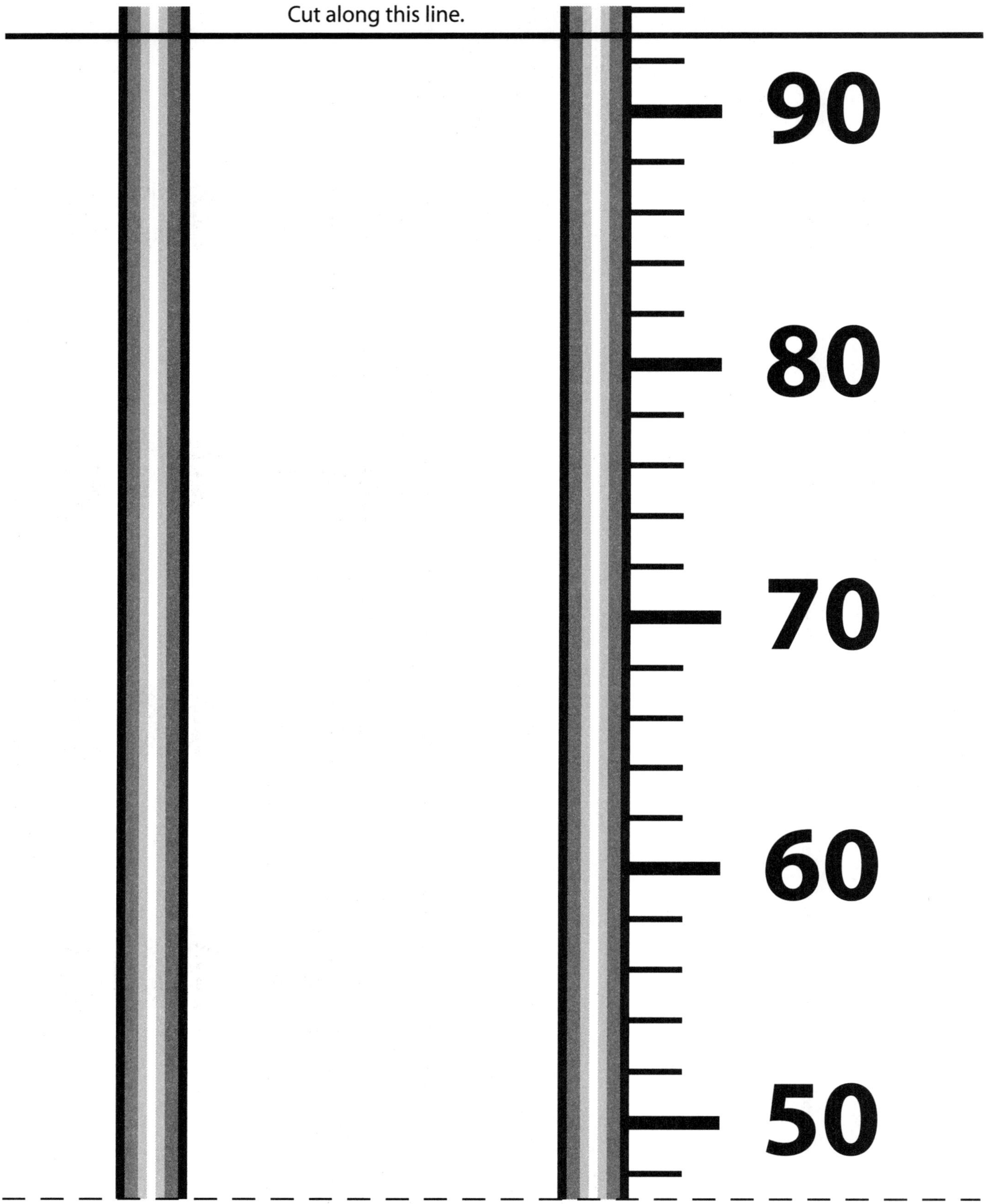

Tape the second page here.

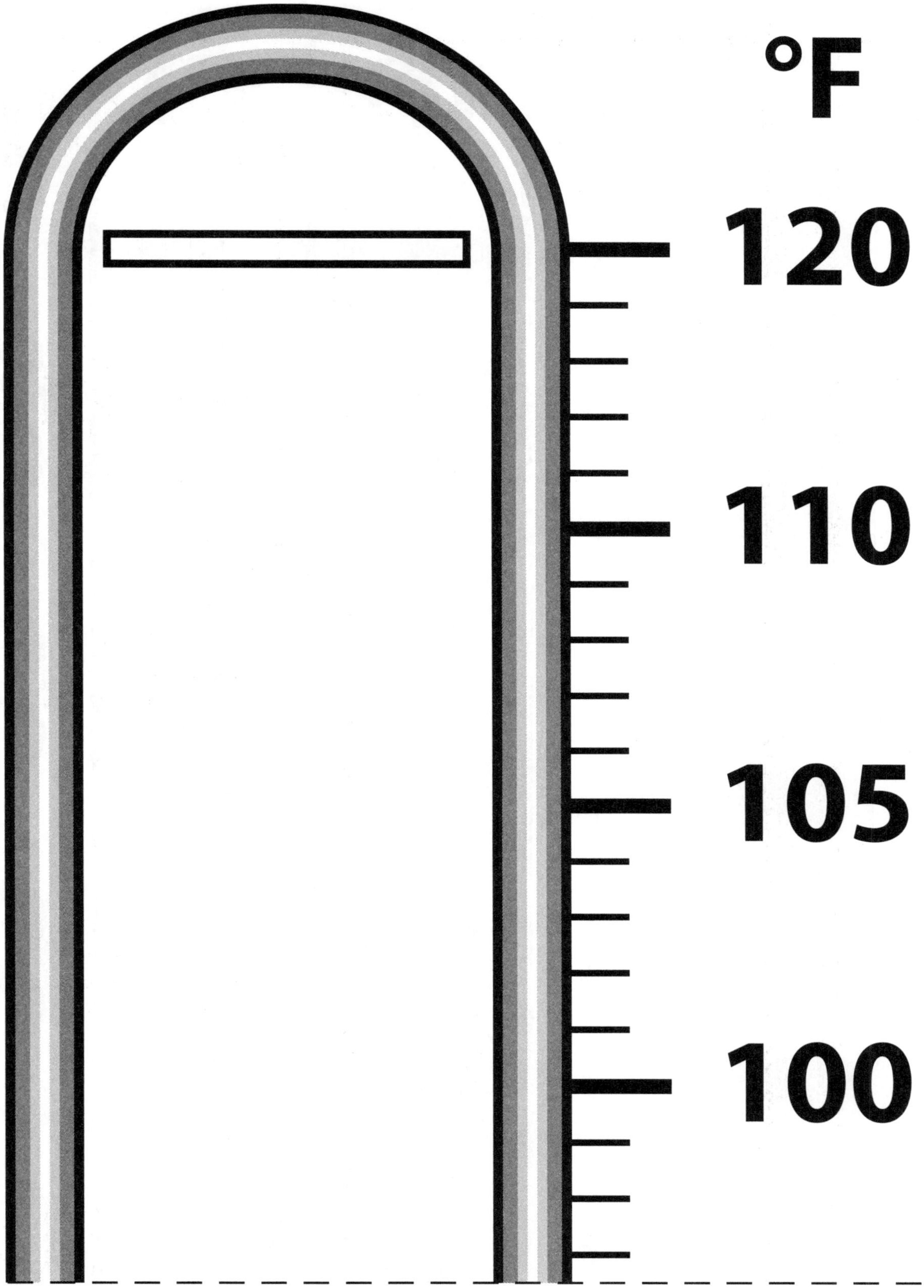

Tape the third page here.

1 2 3 4 5	6 7 8 9 10	11 12 13 14 15	16 17 18 19 20	21 22 23 24 25
TAB	TAB	TAB	TAB	TAB

5	10	15	20	25
4	9	14	19	24
3	8	13	18	23
2	7	12	17	22
1	6	11	16	21
TAB	TAB	TAB	TAB	TAB

A Tool for Temperature

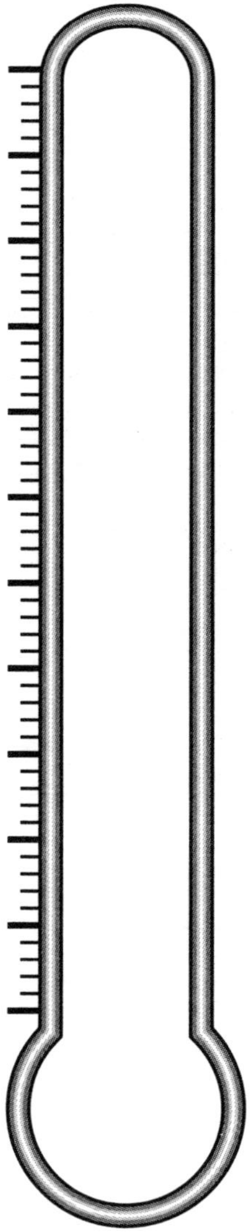

I predict ______ °

It is ______ °

Classroom temperature

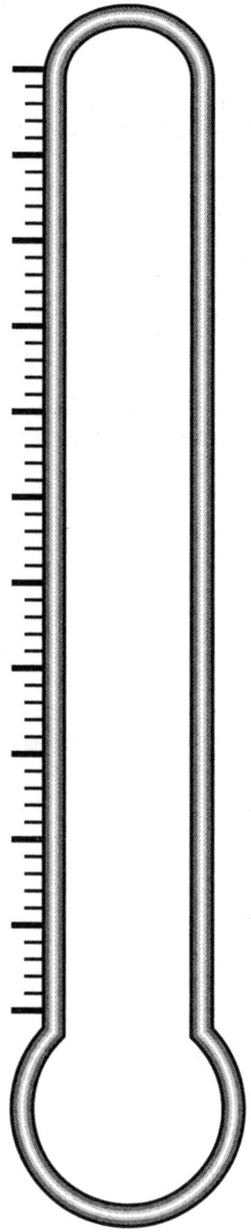

I predict ______ °

It is ______ °

In the shade

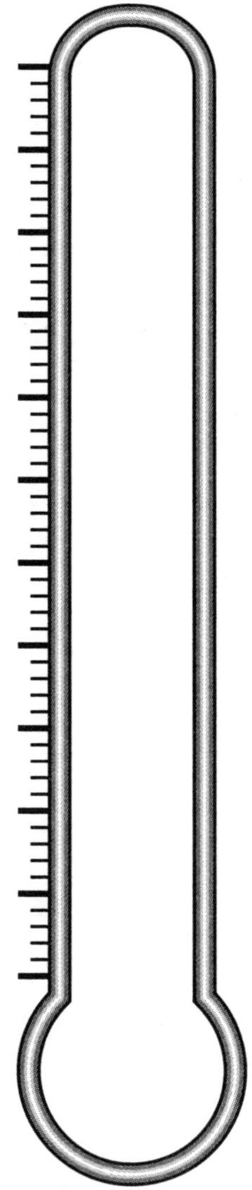

I predict ______ °

It is ______ °

In the sun

Watching the Weather

Topic
Observing the weather

Key Question
What is the weather like today?

Learning Goal
Students will observe and record weather conditions over a long period of time.

Guiding Documents
Project 2061 Benchmarks

- *Describing things as accurately as possible is important in science because it enables people to compare their observations with those of others.*
- *Some events in nature have a repeating pattern. The weather changes some from day to day, but things such as temperature and rain (or snow) tend to be high, low, or medium in the same months every year.*
- *Simple graphs can help to tell about observations.*
- *Some things are more likely to happen than others. Some events can be predicted well and some cannot. Sometimes people aren't sure what will happen because they don't know everything that might be having an effect.*

NRC Standards

- *Weather changes from day to day and over the seasons. Weather can be described by measurable quantities, such as temperature, wind direction and speed, and precipitation.*
- *The sun, moon, stars, clouds, birds, and airplanes all have properties, locations, and movements that can be observed and described.*

*Common Core Standards for Math**

- *Classify objects and count the number of objects in in each category. (K.MD)*
- *Represent and interpret data. (1.MD/2.MD/3.MD)*

Math
Counting
Graphing

Science
Earth science
weather

Integrated Processes
Observing
Comparing and contrasting
Predicting
Inferring
Collecting and recording data

Materials
Weather symbols (see *Management 1)*
Weather calendar chart (see *Management 2)*
Crayons or colored pencils
Student pages

Background Information
Weather affects our lives in many ways. The type of clothing we wear depends largely on the weather. Many businesses are affected by weather, good and bad. Farmers need rain to water crops and sunshine to make them grow. They need clear weather to plant and harvest crops. Rain, snow, fog, and wind can disrupt transportation. Inclement weather even affects children's plans for outdoor recreation.

Weather is the condition of the atmosphere at a certain time in regards to the amount of sunlight, precipitation, clouds, and range of temperature. The weather can be windy, rainy, cold, hot, sunny, or a combination of things.

Weather is different in different locations at any given time. Oceans, mountains, and wind patterns create various weather systems. Young children tend to think that the weather they experience today is the same everywhere. It is a challenge to help them realize that weather varies greatly from region to region.

Management

1. Print multiple copies of the large weather symbols for the weather calendar chart, cut them apart, and keep them in small boxes or plastic bags.
2. Make a large weather calendar chart on poster board or with yarn on a bulletin board.

Procedure

1. Begin the discussion by asking the students to describe the day's weather. Go outside if necessary. Have them describe what the weather was like yesterday. Ask them how the weather affects them.

2. Distribute the first student page, *Observe the Weather.* Tell the students to use their senses to observe the weather.
3. Discuss how they or their parents decided what the weather would be like today. For example: Did they look out the window to see if it was sunny, raining, snowing, etc., or did they listen to the weather forecast?
4. Brainstorm words to describe all sorts of weather. (foggy, rainy, snowy, cloudy) List these on the board or on a large cloud-shaped piece of bulletin board paper.
5. Explain that weather forecasters keep records of daily weather to help them predict the weather for the following days. Tell the students that for the next month, the class will use some weather symbols to make a record of the weather on a class calendar chart. Show and explain each of the symbols.
6. At the same time each day, ask the students which symbol should be posted to best represent the day's weather. Post the day's symbol on the class calendar chart.
7. After a few days of observing, ask the students to predict what they think the weather will be like tomorrow. Have the class vote and post the most frequent guesses next to the chart. Compare their predictions with the actual weather conditions on the following day.
8. On the graph page, have students record the number of days that each type of weather occurred. Keep records for more than one month (or season) and compare the number of clear days, rainy days, foggy days, etc.
9. Have students record the week's weather from the class calendar onto their chart page. They can draw or write their observations on the chart.
10. Use the *My Favorite Weather* page as a culmination for weather observation.

Connecting Learning

1. What is a word you would use to describe the weather today?
2. How many days did we keep a record of the weather?
3. How many clear days did we have during the month? How many days during the month were cloudy? Did we have more clear days than cloudy? How do you know?
4. Looking at the graph, what kind of weather did we have the most?
5. Did we have any rain (or snow) during the month? If so, how many days?
6. What was the weather like today? What was it like a week ago? Why is it easier to tell what it was like a week ago than to predict what it will be next week?
7. What type of clothing would you wear for a hot, sunny day? What would you wear on a cold, wet day? How does clothing color, weight, and thickness differ with the kind of weather we are having?
8. What kind of weather do you like the best? Why? What do you do on those days?
9. What are you wondering now?

Extensions

1. Weather affects all life on Earth. Discuss the danger of electrical storms and procedures to stay safe during dangerous weather. Make a class plan for any severe weather in your area.
2. Make some paper dolls, then make clothing to coordinate with the local weather. Tell the students they are to pick the clothes for the dolls, dress them according to the day's weather and post them on the board near the weather calendar for the day.

Curriculum Correlation

Art

The students can create pictures to match the weather outside by painting bright sunny days with bold tempera colors, rainy days with watercolor washes, etc.

Home Links

1. Send home a note asking parents to sit with their child and listen to the weather report (TV, radio, or newspaper). Discuss the weather predictions the following morning in class.
2. Have the students make a home plan with their families for severe weather.

Watching the Weather

Observe the Weather

Use your senses to observe weather. Draw and write descriptions.

See	**Hear**
Smell	**Feel**

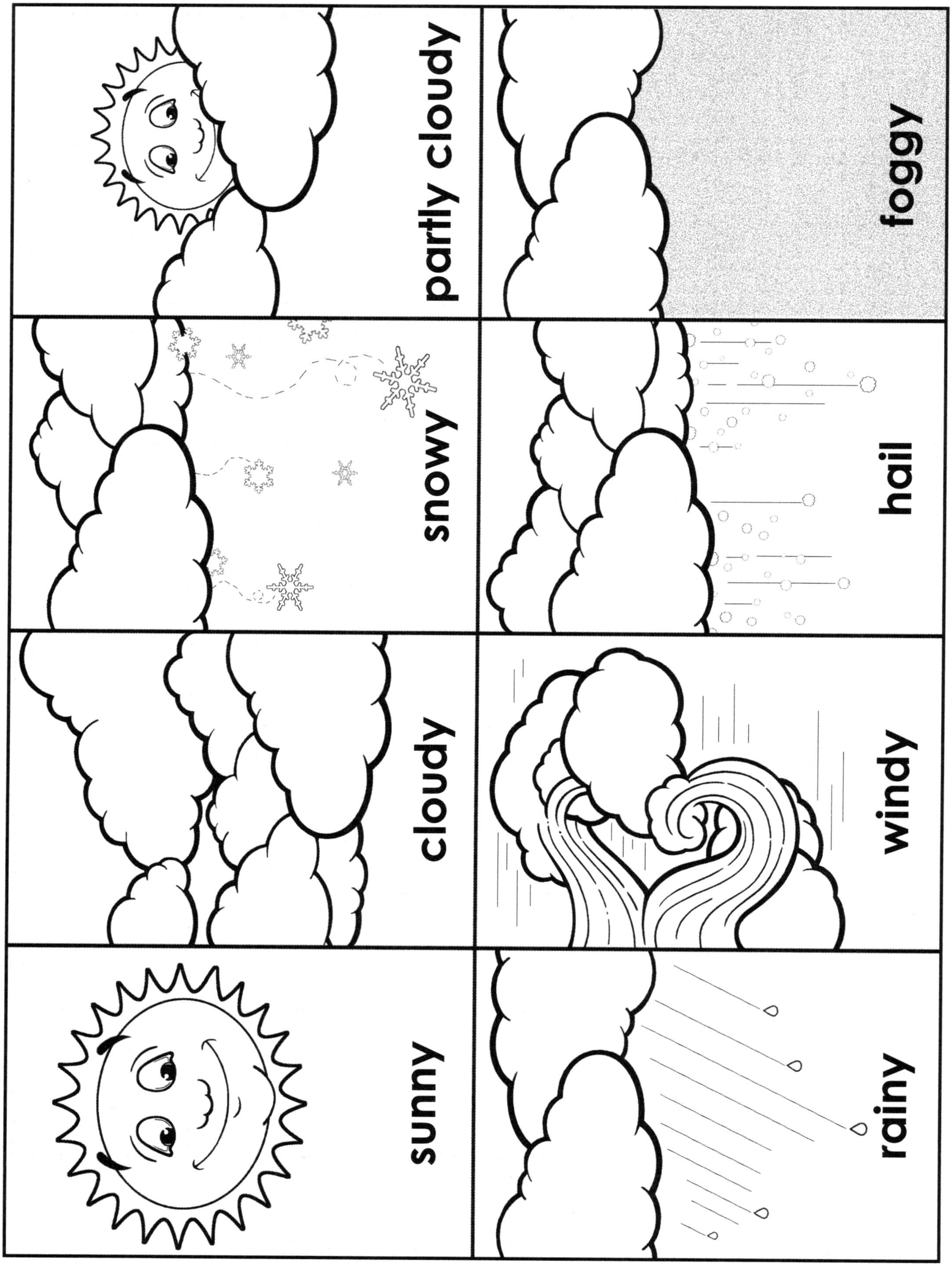

sunny
cloudy
snowy
partly cloudy
rainy
windy
hail
foggy

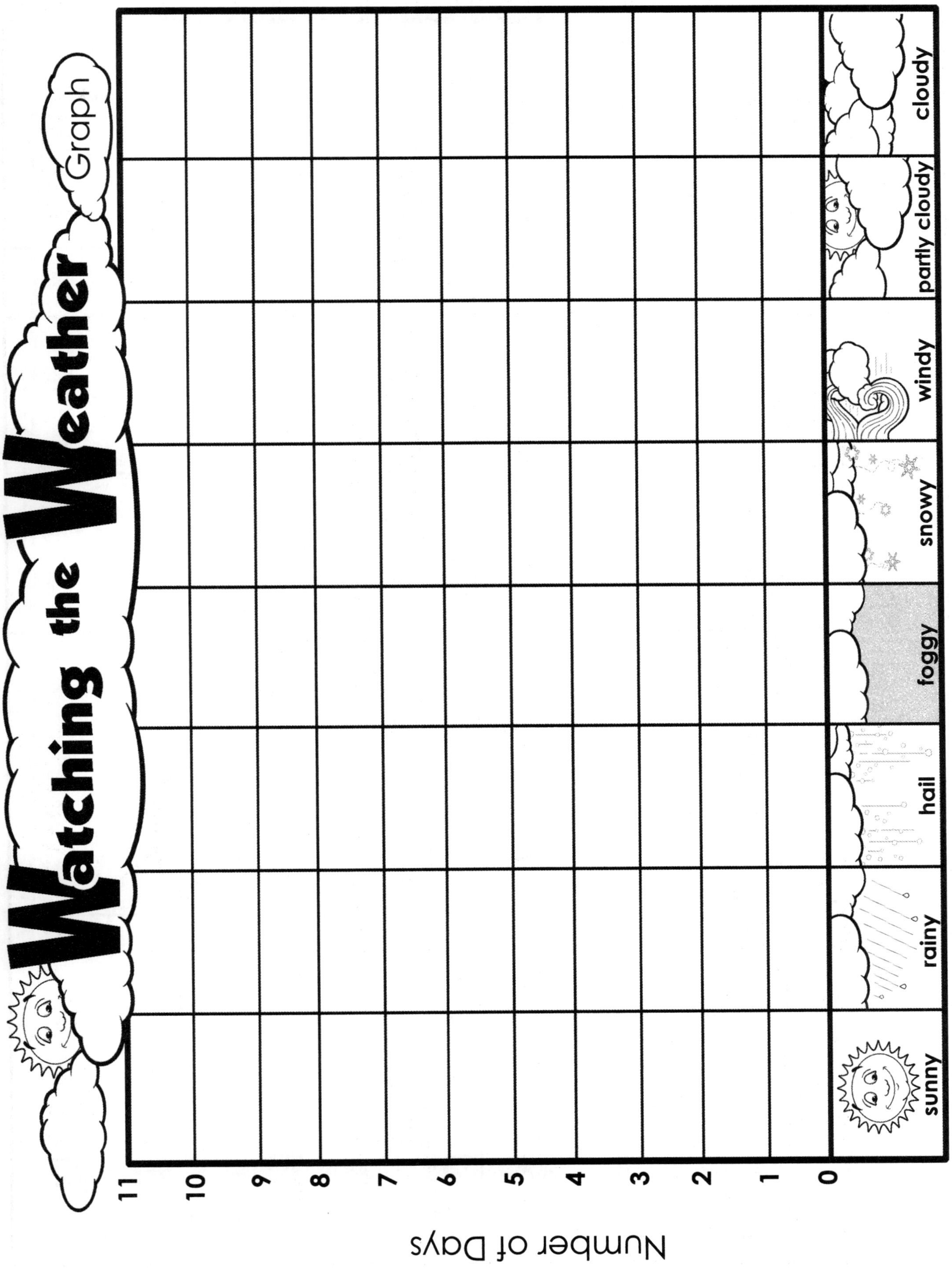

Watching the Weather
Graph
sunny
rainy
hail
foggy
snowy
windy
partly cloudy
cloudy
0
1
2
3
4
5
6
7
8
9
10
11
Number of Days

Watching the Weather Chart

	Monday	Tuesday	Wednesday	Thursday	Friday
Week One					
Week Two					
Week Three					
Week Four					

Watching the Weather

My Favorite Weather

1. Draw a picture of your favorite type of weather. Put yourself in the picture.

2. Write words to describe your favorite weather.

Dress for the Weather

Topic
Weather changes

Key Question
How does the weather affect what we wear?

Learning Goals
Students will:
- observe and describe seasonal changes in weather, and
- decide how they should dress based on the weather.

Guiding Documents
Project 2061 Benchmarks
- *Change is something that happens to many things.*
- *Simple graphs can help to tell about observations.*
- *Some events in nature have a repeating pattern. The weather changes some from day to day, but things such as temperature and rain (or snow) tend to be high, low, or medium in the same months every year.*

NRC Standard
- *Weather changes from day to day and over the seasons. Weather can be described by measurable quantities, such as temperature, wind direction and speed, and precipitation.*

Science
Earth science
weather

Integrated Processes
Observing
Collecting and recording data
Interpreting data
Drawing conclusions
Generalizing

Materials
Set of *Dress for the Weather Cards*
Student *Dress for the Weather* book
Dress for the Weather Calendar
Camera (see *Management 3)*
Stapler
Scissors
Weather wear

Background Information

Most young children are unaware that people wear clothing for protection and adornment. Often students don't realize that what they wear is influenced by the weather. They just know that their parents make them wear something. They are often not aware that they have to wear a coat and long pants because it is winter and that they are allowed to wear shorts in the summer when it is warm. They usually become aware of what they are wearing when it is uncomfortable for one reason or another. For example, they want to wear shorts and no coat if they are warm. When they get cold, they want to put on a sweater.

In this activity, students will become aware of the influence that weather has on their clothing choices through discussion and charting of the daily weather. They will then be asked to choose an appropriate outfit for the day from their *Dress for the Weather* books.

Management
1. Make multiple copies of the *Dress for the Weather Cards* so that each child can glue down the picture to keep a record of the weather for one month.
2. Make one copy of the student *Dress for the Weather* book and calendar for each child.
3. Prior to teaching this lesson, take pictures of the students in the class wearing different weather gear. For example, shorts and a t-shirt, raincoat and boots, winter coat and long pants, etc. Note—when taking pictures of the children in your class, it is best to get parents' permission.
4. Plan to display the pages from the *Dress for the Weather* book using a projection device.

Procedure
1. Engage the students in a discussion about the weather. Ask the class to describe the day's weather. If necessary, take the students outside to observe the weather.
2. Discuss how parents decide what the weather is going to be like today. [listen to the weather on television or radio, go outside to check the weather, etc.]
3. Ask the class to help you generate a list of words that are often used to describe the weather. [foggy, rainy, snowy, cloudy, sunny] Record these on the board or chart paper.
4. Ask the students how today's weather influenced what they are wearing.

5. Display the photos. Discuss the various outfits in the photos. Ask the students for what type of weather each of the outfits would be appropriate.
6. Tell the class that each day for the next several weeks they will be keeping a record of the weather. Explain that each day they will observe the weather, choose the weather card that best illustrates the day's weather, and glue the weather card onto the *Dress for the Weather* calendar.
7. Display the winter jacket, boots, and long pants from the *Dress for the Weather* bookusing a projection device. Ask the class what the weather would most likely be if they wore that outfit to school. Repeat several times using different clothing combinations.
8. When the students are familiar with the various clothing combinations, show them an assembled copy of the student book and demonstrate how you can flip the pages to reveal various combinations appropriate for different kinds of weather. Explain that the students will be making their own copies of the book and that each day as they record on their calendars, they will also use their flip books to show the appropriate clothing to wear in that type of weather.
9. Distribute scissors and copies of the student book to each child. Assist the students with the assembly of their books. Ask the students to draw facial features and hair on the character.
10. When books are assembled, practice using the books by asking the class to show you what they should wear on a rainy day. Have the students compare their choice with their neighbors. Continue this procedure each day for a month.

Connecting Learning

1. What is the weather like today? How did you dress?
2. What should you wear when it is very hot outside?
3. What should you wear when it is very cold outside?
4. Why shouldn't you wear shorts on a cold winter day?
5. Other than weather, what influences what we wear?
6. How might our *Dress for the Weather Calendar* be different in the summer? ...winter?
7. What are you wondering now?

Extension

Have the class sort the pictures by season, hot or cold temperature, etc.

Dress for the Weather

by

____

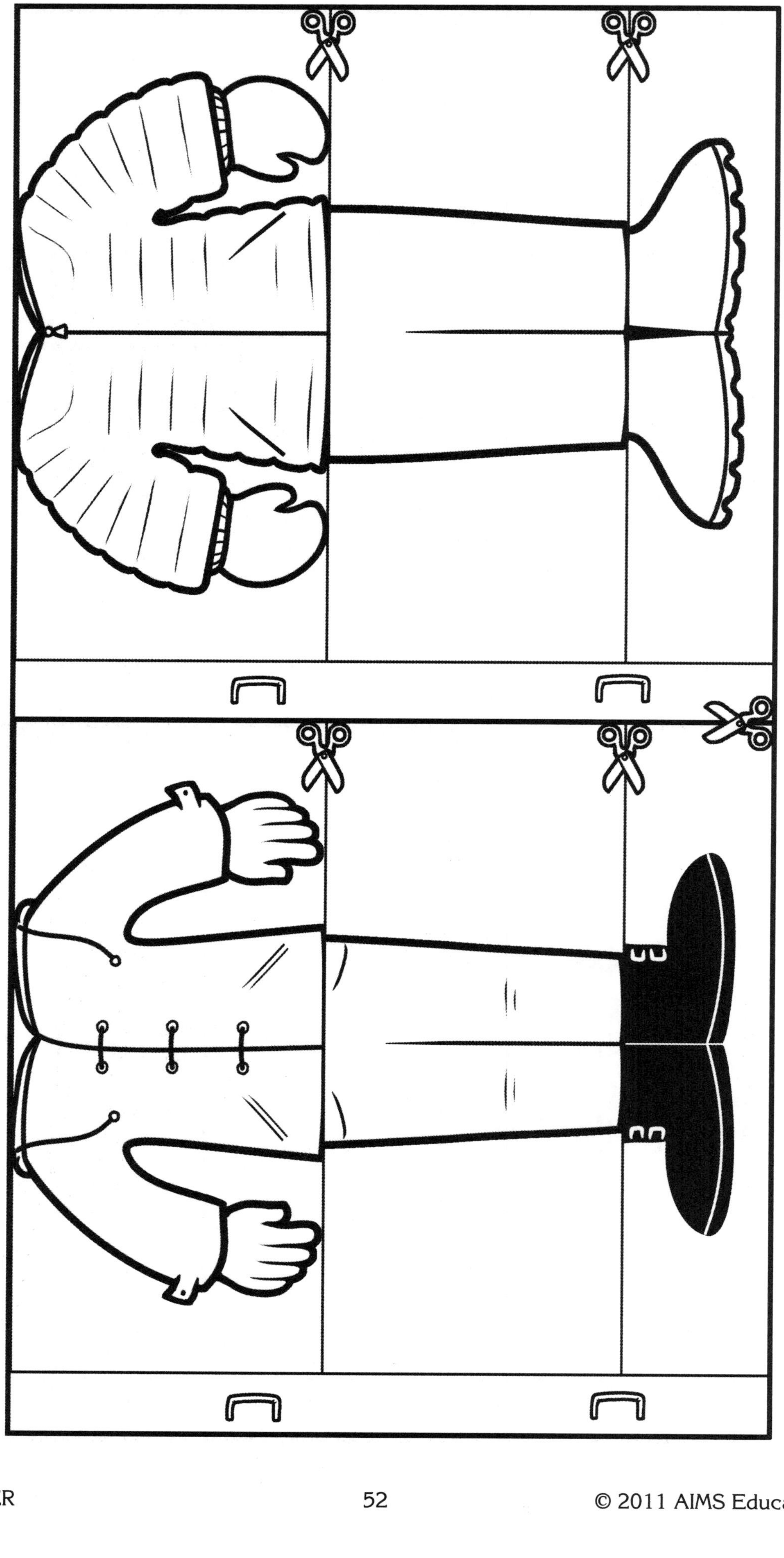

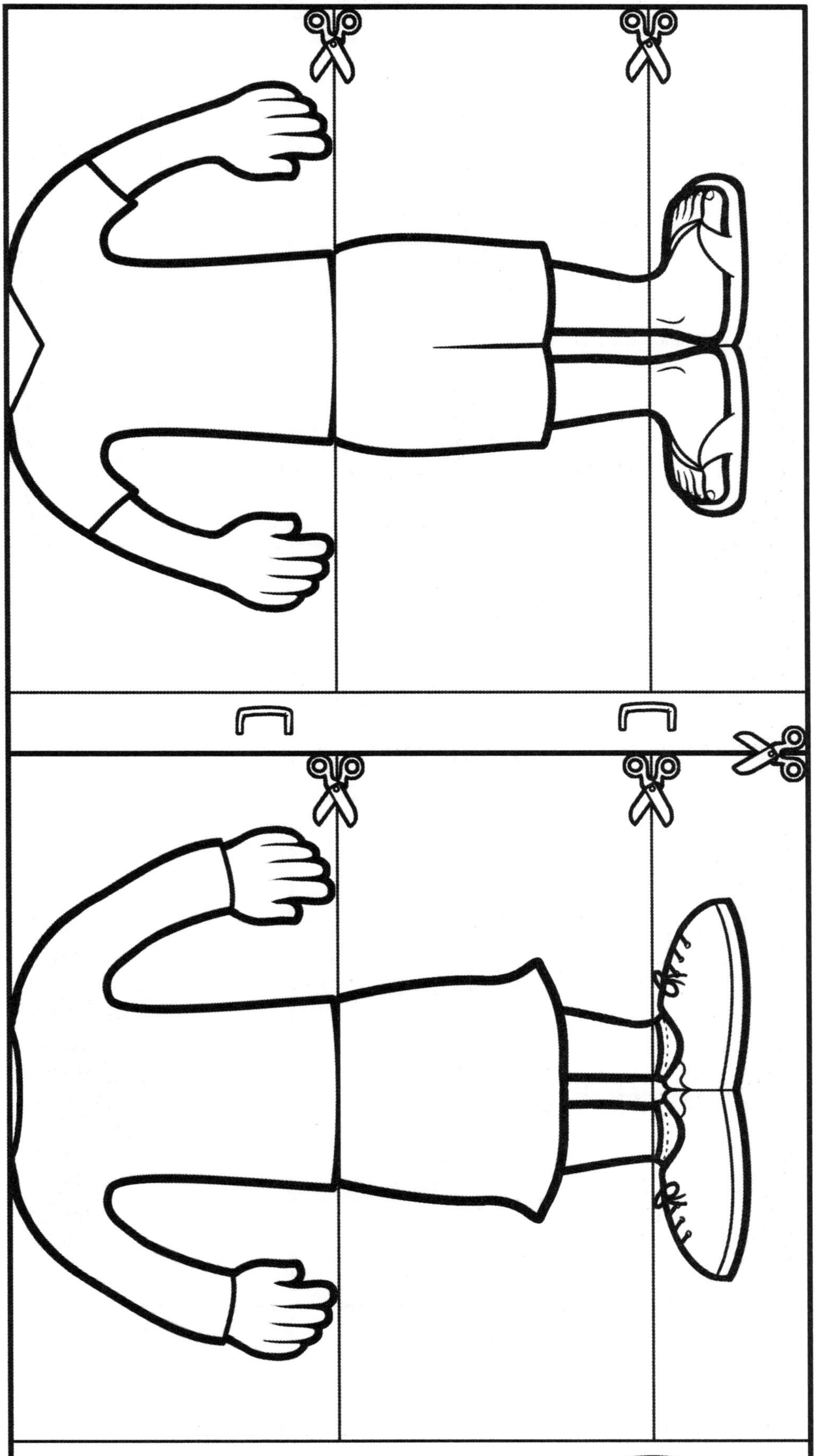

Dress for the Weather
Calendar

	Monday	Tuesday	Wednesday	Thursday	Friday
Week 1					
Week 2					
Week 3					
Week 4					

Dress for the Weather
Cards

Weather Changes

Tune: Clementine

Wea-ther chan-ges, wea-ther chan-ges, wea-ther

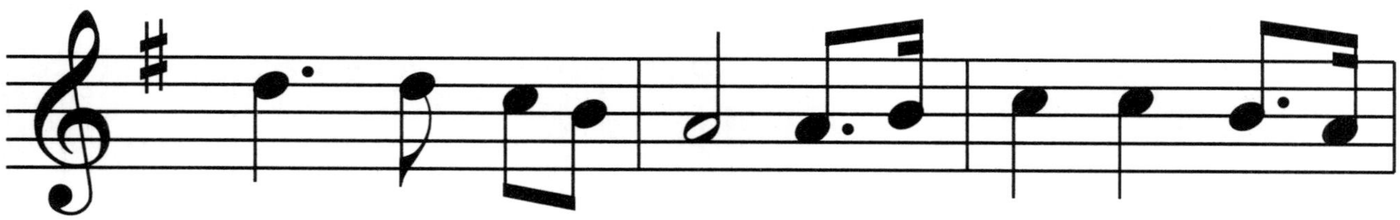

chan-ges ev-ery day. Ev-ery day the wea-ther

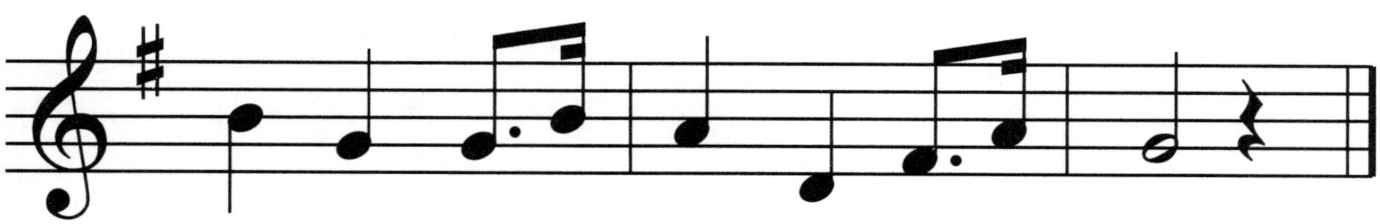

chan-ges, wea-ther chan-ges ev-ery day.

First it's cloudy, then it's rainy;
Weather changes everyday.

After snow it can be sunny,
Melting all the snow away.

In the morning, it is chilly.
I wore mittens here today.
But by recess, it is warm out;
Weather changes everyday.

Weather changes through the seasons,
But each season is the same;
Spring and summer, fall and winter,
We know each one by its name.

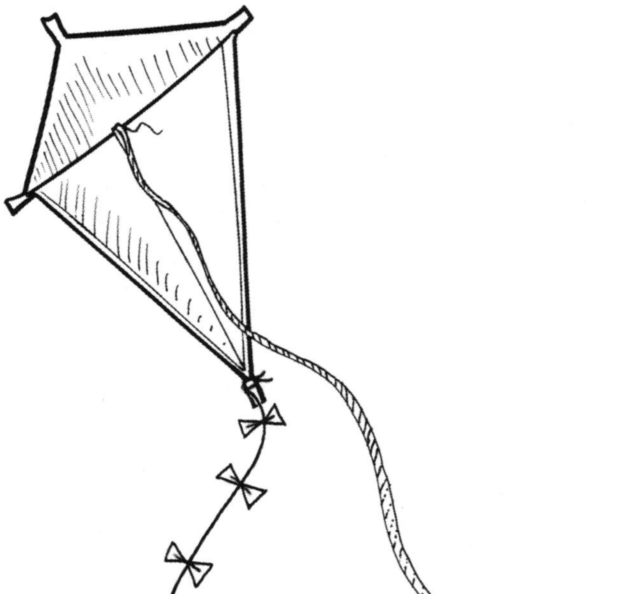

In the summer, I go swimming,
But in fall, it's cold at night.
In the winter, we make snowmen.
In the spring, I fly a kite.

Weather changes with the seasons.
Weather changes day to day,
But we watch for all the changes
To prepare in every way.

Weather Windows

Topic
Weather

Key Question
How does the weather change from season to season?

Learning Goal
Student will illustrate the weather conditions of each of the four seasons.

Guiding Documents
Project 2061 Benchmarks
- *Describing things as accurately as possible is important in science because it enables people to compare their observations with those of others.*
- *Some events in nature have a repeating pattern. The weather changes some from day to day, but things such as temperature and rain (or snow) tend to be high, low, or medium in the same months every year.*
- *Some things are more likely to happen than others. Some events can be predicted well and some cannot. Sometimes people aren't sure what will happen because they don't know everything that might be having an effect.*

NRC Standard
- *Weather changes from day to day and over the seasons. Weather can be described by measurable quantities, such as temperature, wind direction and speed, and precipitation.*

Science
Earth science
  weather
  seasons

Integrated Processes
Observing
Communicating
Predicting
Comparing and contrasting
Generalizing
Relating

Materials
Scissors
Crayons
Chart paper
My Weather Window booklet, one per student
Cloth scraps, optional
Glue, optional

Background Information
Weather is the state of the atmosphere at a given time and place with regard to temperature, moisture, and barometric pressure. It changes from season to season and varies by location. The weather affects our lives in many ways. It determines the clothing we wear and the activities in which we participate. It affects our mode of transportation and often influences the foods that we eat.

By illustrating weather conditions that exist during the four seasons, students will demonstrate their understanding of the changes that take place from season to season. Because young children's understanding of seasonal changes is generally limited to their experiences, it is important for them to be exposed to information about seasonal changes that occur in other parts of the country. This can be done through the use of books and/or Internet daily weather reports from various places in the country.

Management
1. Each student will need one copy of the *My Weather Window* booklet. Prior to distributing the booklets, cut out the window and laminate. Curtains can be added to the cover by gluing or stapling fabric scraps along the top edges of the window, and small pieces of ribbon or yarn can be used as curtain tiebacks.

Procedure
1. Have the class describe what the weather was like when they arrived at school. Ask the students if they think the weather has changed during the day. Take the class outside and observe any changes. Have the students compare today's weather with yesterday's weather. Discuss how weather changes from moment to moment, day to day, and season to season.

2. Discuss how the weather affects what they wear and what activities they are able to participate in. For example, if it is raining, they will have indoor recess; if it is sunny, they will be able to play outside. If they are inside for recess, they can play checkers, heads up/seven-up, etc. If they go outside, they can play ball, tag, etc.
3. Have the students name the four seasons. Write the name of each season on a piece of chart paper, leaving room to record students' responses. Ask them to describe what the weather is like during each season. Record their responses under the appropriate season on the chart paper.
4. Ask the class what types of activities they participate in during each of the four seasons. For example: winter—skiing, sledding, snowball fights, staying inside due to snow or rain, etc. Record student responses under the appropriate season.
5. Through the use of books and/or the Internet, find out what the seasons are like in different parts of the country. For example, Hawaii's average temperature in January is 71 degrees while Alaska's average temperature is 10 below zero during the month of January. This means that while kids in Alaska are playing in the snow, kids in Hawaii might be swimming in the ocean.
6. Give each student a copy of the *My Weather Window* booklet, crayons, and scissors. Direct the students to fold the cover and cut the other pages apart.
7. Have the students locate the page with the current season label at the top. Discuss what kind of weather we might see during the current season. Instruct the students to draw a scene that illustrates what the weather looks like during this particular season. Have them use words from the chart paper to describe that season on the lines at the bottom of the page. Instruct them to illustrate and describe the other three seasons when they have completed the current season. If material is available, they can also make curtains for the book covers at this time.
8. When all illustrations and descriptions are finished and students are ready to assemble their booklets, question them about the order of the seasons. Ask what season comes before and directly after the current season. Assist students in assembling their weather window booklets in the correct order, starting with the current season.
9. Close with a time for students to share their weather windows.

Connecting Learning

1. What season are we in now? How would you describe the weather today?
2. During what season does cold weather settle in?
3. How are spring and fall weather conditions different? How are they alike?
4. What does summer look like in your weather window? How is it different from your winter window scene?
5. What are you wondering now?

Fold

WINTER

In winter...

SUMMER

In summer...

SPRING

In spring…

FALL

In fall…

A Snap of Time

Topic
Seasonal changes in trees

Key Question
How does a tree change during the school year?

Learning Goal
Students will adopt a tree for the school year and observe the changes in various aspects of the tree.

Guiding Document
Project 2061 Benchmarks
- *Change is something that happens to many things.*
- *Things change in some ways and stay the same in some ways.*

Science
Earth science
 seasons
Life science
 seasonal changes

Integrated Processes
Observing
Comparing and contrasting
Communicating
Collecting and recording data
Predicting
Generalizing

Materials
Clipboards or books and clothespins
Yarn
Hand lenses
The Seasons of Arnold's Apple Tree by Gail Gibbons
Student pages

Background Information
Watching a tree can give us a glimpse of seasonal changes. Most broadleaf trees lose their leaves in the fall when the amount of sunlight diminishes and the days become cooler. In contrast, conifers lose their needles only every three to four years. However, even with conifers, seasonal changes can be seen below the tree.

It is important for the observer to look for seasonal changes below and around the tree as well as looking at the changing tree. The soil and soil covering below the tree can change and therefore support us in noticing seasonal changes.

In the spring, buds are visible on the tree and we begin to see new leaves. At that time, we will most likely see new grass and small flowers, and the soil will feel warmer under the tree. While the tree remains green for most of the summer, the soil and plant life under it may start to dry. In fall and winter, leaves of a deciduous tree will change color and drop, covering the ground. In the winter the tree may be bare, and the grass and plant life below the tree may be dormant. Winter may bring frost or snow with the ground becoming cold and unpenetrable. Dead leaves may cover the ground. Because a tree can be a life support system for various animals, the animals associated with the tree may change as the seasons change.

Management
1. It will be very important that the adopted trees show seasonal changes during the school year. If a student is adopting a tree in his or her neighborhood, send a letter to parents encouraging them to help select a tree that will change during the seasons and be easily observable all year long. If the trees being adopted are on or near the school grounds, walk the area and select a spot that will provide enough trees for each pair of students.
2. If this activity is done at school, it is best done in partners.
3. This activity will be most effective if it is started at the very first of the year so students will see the most change in their trees. At least once during each season, have the students do a formal observation.
4. Field recording can be difficult. The use of clipboards or books with clothespins used to keep the paper in place can make outdoor recording easier.
5. Hand lenses (item number 1977) are available from AIMS.

Procedure
1. Read *The Seasons of Arnold's Apple Tree* by Gail Gibbons. Tell the students they too are going to watch a tree for the school year. Ask them what they know about trees and how they change.
2. Take students on a walk to observe trees on the school grounds and in the neighborhood. If possible, include some information during your walk about trees that obviously change and those that do not. If they are going to adopt trees at school, have partners select their trees and sketch them. A strategy that will help them observe and sketch better is to look a little, sketch a little; look a little, sketch a little. When students are done, they

should trade pictures with their partners and give each other suggestions about possible additions to their sketches.

3. Wait only a day or two and visit the trees again. This time make the close-ups sketches and collect data in relation to things they might see *on* or *under* the bark, *in* and *on* the branches and leaves, and *around* the roots and Earth *below*. Have students touch the ground and discuss the warmth or coolness of the soil, as well as observe and make a record of the things growing or laying on the soil. They should look carefully at the bark around the base of the tree.
4. When students visit the tree the second time, have each partner select a branch and tie a piece of colorful yarn on a branch. Tell them they should look very carefully at the branch and its leaves each time they observe their tree. Encourage them to count the leaves, describe the color, smell the leaves and branches, and touch them.
5. While they are sketching, have the students also look for any living things that may be using the tree. It is important that a hand lens be used when looking at the tree bark, leaves, and critters. Have the students make a quick sketch of any critters.
6. Encourage a "silent observation" period while the students are visiting the tree. Direct students to sit near their trees and to remain quiet for a least one minute.
7. After the silent observation, allow time for students to share what they observed. Discuss the fact that looking is only one way to learn about their tree; listening, smelling, and touching also give us much information.
8. Each time the students visit their trees, continue to encourage them to sketch each part of the tree and its immediate environment as well as to make field notes of their observations. (Younger learners can dictate their observations as well as draw them.)
9. If appropriate, students can collect fallen leaves or twigs to include in their albums of their adopted trees.
10. To obtain a print of the leaf or bark, have the students do a rubbing.

Connecting Learning

1. How has your tree changed?
2. What did your tree look like in the summer? ...fall? ...winter? ...spring?
3. Describe the changes in the branches (bark, ground, inhabitants, etc.) of your tree throughout the seasons.
4. How are the changes related to the seasons?
5. What kind of things lived in, on, and above the tree? Were any in all three places? Were these things there in all of the seasons? Why or why not?
6. How did having a partner in this project help you learn about your tree?
7. What was your favorite season to observe your tree? What made it your favorite season?
8. What are you wondering now?

Extensions

1. Invite an amateur photography group or the high school photography class to take pictures of the students and their trees several times during the year.
2. Invite a local forester or visit a local nursery to learn more about trees.
3. Write the National Arbor Day Foundation for information:
   The National Arbor Day Foundation
   100 Arbor Avenue
   Nebraska City, NE 68410
4. Use the Internet to write to other schools and invite them to do this activity and share their observations so comparisons can be made.The ePals website (www.epals.com) is a good one for finding other teachers who are interested in collaboration.

Curriculum Correlation

Gibbons, Gail. *The Seasons of Arnold's Apple Tree.* Harcourt Brace Jovanovich. New York. 1984.

My tree in the Winter

Close up of a branch

Close up of the bark

My tree

The ground under my tree

My tree in the Spring

Close up of a branch

My tree

Close up of the bark

The ground under my tree

My tree in the Summer

Close up of a branch

My tree

Close up of the bark

The ground under my tree

My tree in the Fall

Close up of the bark

Close up of a branch

My tree

The ground under my tree

CERTIFICATE OF ADOPTION

I hereby certify that this ____________ tree has been lovingly adopted by _______________ on this __ day of __________ in the year ______.

Signed,

____

Teacher

____

Adoptive parent

Season-O

Topic
Seasonal changes

Key Question
As the seasons change, what observations of the tree and its surroundings can you make?

Learning Goal
Students will play a game that focuses on seasonal changes in relation to a tree.

Guiding Document
Project 2061 Benchmarks
- *Change is something that happens to many things.*
- *Things change in some ways and stay the same in some ways.*

Science
Earth science
  seasons
Life science
  trees

Integrated Processes
Observing
Comparing and contrasting
Communicating

Materials
Chart paper
Game board and game pieces (see *Management 2, 3*)
Paper lunch bags

Background Information
Seasonal changes in many areas are very dramatic but often are not carefully observed by the young learner. In this activity, they apply and extend their learning about seasonal changes by playing a game.

Management
1. It is important that the students have experienced the activity *A Snap of Time* before playing *Season-O* so they have some understandings about seasonal change.
2. *Season-O* is played much like the game Bingo in which the winner completely fills in a vertical, horizontal, or diagonal line. Copy the game pieces (seasonal markers) and game board (tree) on three different colors of paper so the win lines will be more evident.
3. Each person (if playing in partners) or each team needs a set of game pieces copied in a color different from the opponent's pieces. Each set of pieces should be put in a paper lunch bag. Students will draw the pieces from the lunch bags.
4. Students can play this game in partners or groups of four. You can also enlarge the game board and game pieces so the whole class can play (see *Procedure)*. The enlarged game board should be placed on a bulletin board where the class can see it. Put a pin in the top center of each grid of the game board. Each team (half of the class) will need a different colored set of game pieces. These game pieces should have a hole punched in the top center of the piece so they can be placed easily over the pin on the game board. Each half of the class will take turns drawing a game piece and putting it on the game board as well as describing what season the game piece depicts and justifying their selection. The first team (half of the class) that gets their color pieces in a horizontal, vertical, or diagonal row wins.
5. Enlarge the book *My Tree, My Tree, All Year Long* to make a big book.

Procedure
1. Motivate the discussion of seasonal changes by reading *My Tree, My Tree, All Year Long.* After reading the book, ask students what changes they have observed in their tree over the year. Record their observations on chart paper or a board that has been divided into four columns labeled *Summer, Fall, Winter,* and *Spring.*
2. Hold up the game board (or display it using a projection device) and ask what the tree might look like in the winter. Discuss what the pictures might be above, on, and below the tree. Hold up (or display) several game pieces and discuss where they would go on the tree. Encourage students to use descriptive language that helps them visualize the tree in the different seasons.
3. Demonstrate how they must both verbalize why the game piece should fit in a certain season's column as well as use the clues from the pictures of the tree to explain the placement of the game pieces.
4. Inform the students that the game is won by getting pieces in the correct places in a vertical, horizontal, or diagonal row. It is important that these terms be discussed and understood.
5. Model how to play the game by gathering the students around a game board. Do this modeling using a student as your partner.

6. Distribute game pieces, paper bags, and game boards to partners or teams.
7. Allow time for students to play *Season-O.*
8. As a class, close with a discussion about how a tree changes through the seasons.
9. After the game has been played several times, place two sets of laminated game pieces and a game board in a box or tub so students may continue to play the game during free time.

Connecting Learning

1. Describe the differences in a tree through the seasons.
2. What parts of the tree seem to change a lot? Which change very little?
3. Where are the biggest changes in a tree observed?
4. Which is your favorite season of the year? Why? What does the tree look like during that time of the year?
5. What are you wondering now?

Extension

Allow students to change the rules to the game and share their new rules with the class.

Curriculum Correlation

Language Arts

Have the class act out *My Tree, My Tree, All Year Long.*

Assessment

Have students create their own versions of *My Tree, My Tree, All Year Long.* These versions should include what is happening above, on, and below the tree during all the seasons of the year.

Home Link

Students can take the game home to play with their families by folding the game board and putting game pieces in a plastic bag.

Season-O

Game Pieces

Season-O

Game Board

My Tree, My Tree, All Year Long

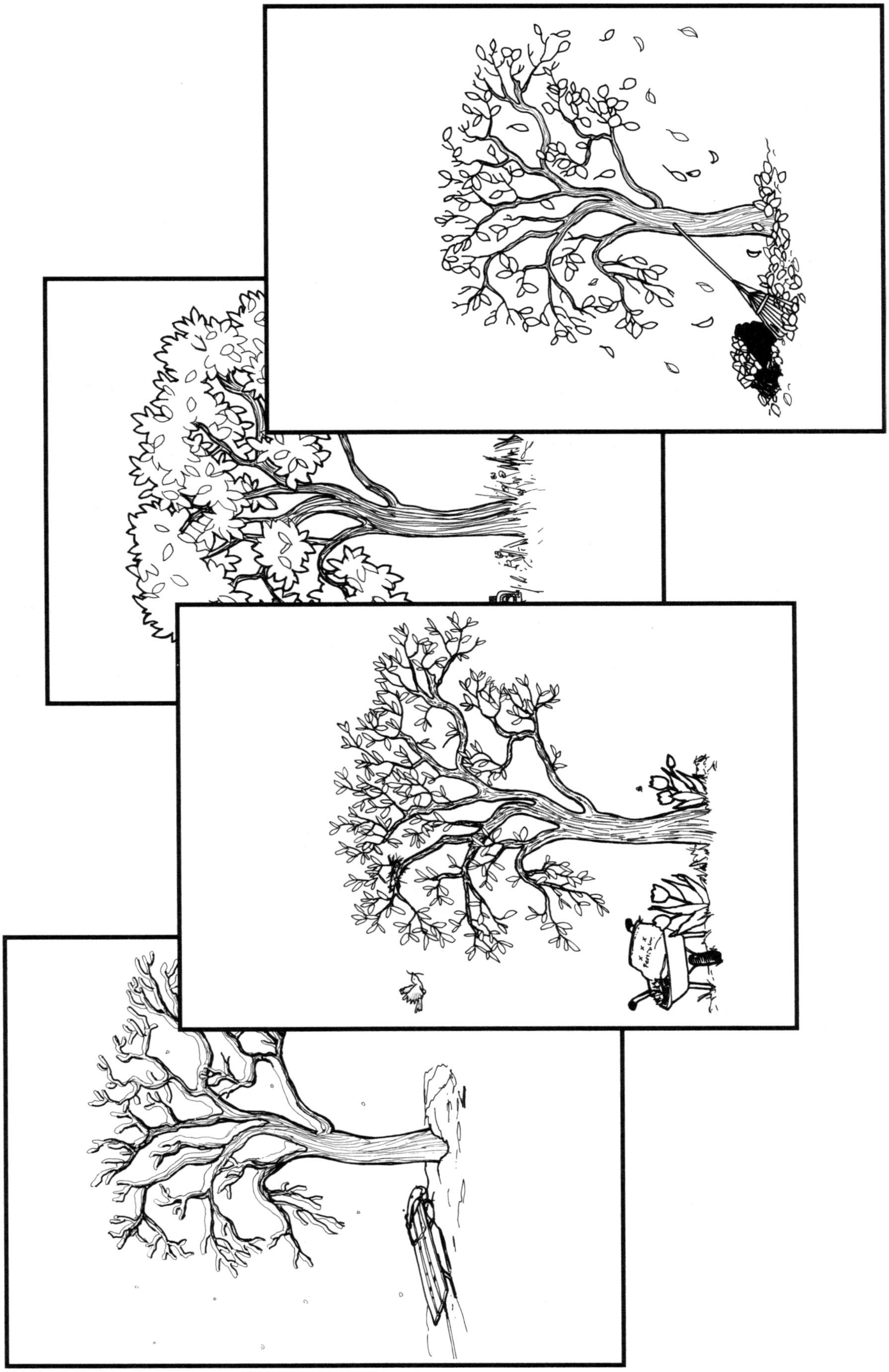

My tree bares its branches in winter;

Snow on the tree,

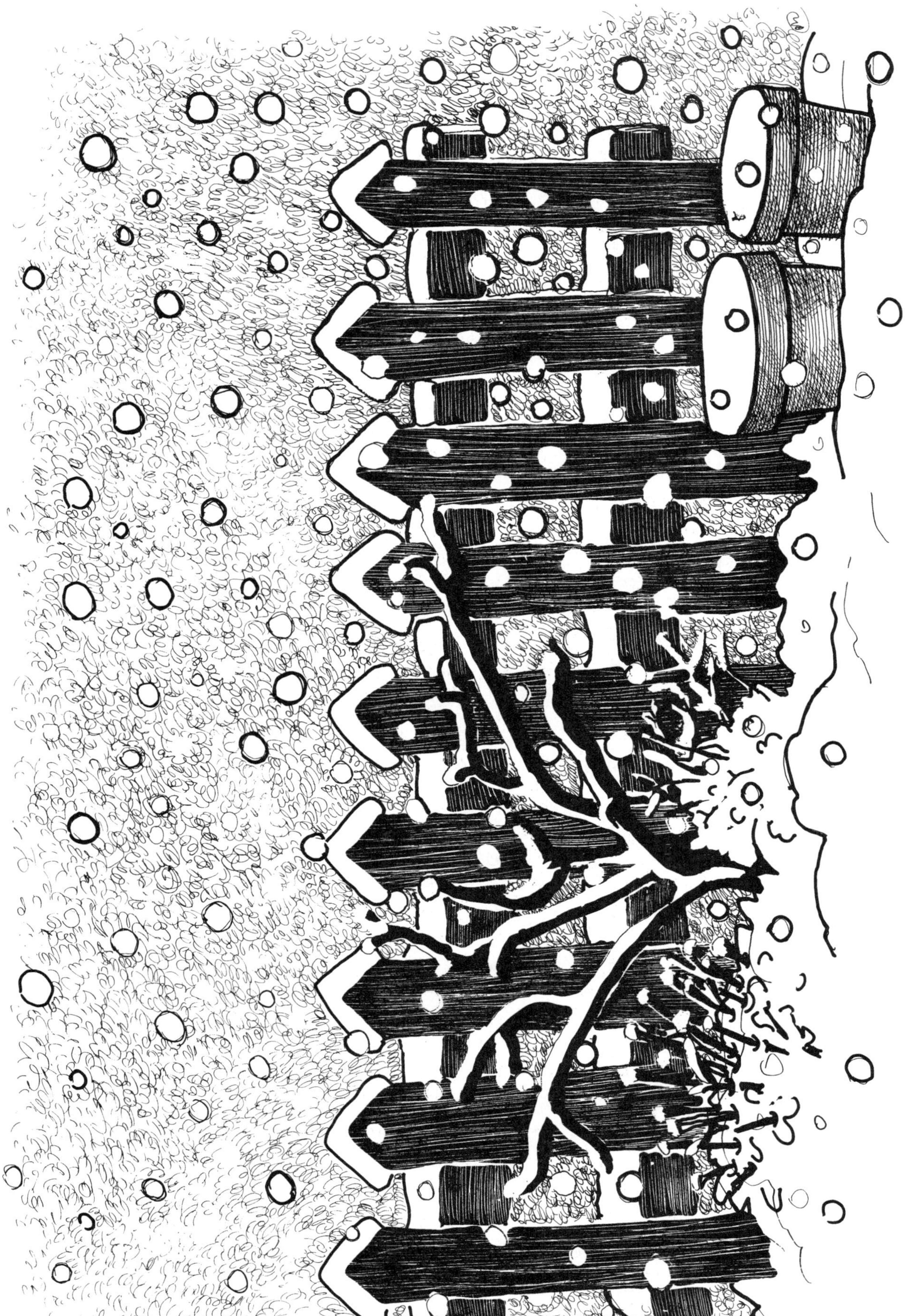

Snow on the flowers,

Snow on me.

My tree blooms and leafs out in springtime;

Rain on the trees,

Rain on the flowers,

Rain on me.

My tree shades and cools in summer;

Sun on the tree,

Sun on the flowers,

Sun on me.

My tree colors and drops leaves in autumn;

Wind on the tree,

Wind on the flowers,

Wind on me.

DIGITAL DIARIES

Topic
Changes

Key Question
How will the tree change over time?

Learning Goal
Students will observe a tree over time and take note of the changes it undergoes.

Guiding Document
Project 2061 Benchmarks
- *Change is something that happens to many things.*
- *Things change in some ways and stay the same in some ways.*

Science
Life science
changes

Integrated Processes
Observing
Comparing and contrasting
Recording data
Communicating

Materials
Digital camera and printer

Background Information
This activity invites children to focus on a tree. They will make careful observations of the tree and then take digital photos of it. Every two to four weeks the children will visit the tree to make fresh observations. They will then compare these observations with the previous observations and photos. This activity can be carried out throughout the year with children noting the seasonal changes of the tree.

Management
1. It is suggested that a deciduous tree be chosen for observations. If desired, choose a deciduous tree and an evergreen for purposes of comparisons.
2. To build vocabulary and reading skills, record observational words on a piece of chart paper. These words can be reviewed at the next observational period.

Procedure
1. Ask children if they have a favorite tree and what it is about that tree that makes it their favorite.
2. Invite children to describe their favorite trees.
3. Ask the children if their favorite trees stay the same all through the year.
4. Tell them that they are going to go outside and make some careful observations of a tree.
5. Have children gather around the tree and share their observations.
6. Take at least four digital photos of the tree, one from each direction.
7. Back in the classroom, record student observational words and post photos.
8. Follow the same procedure every two to four weeks.

Connecting Learning
1. What do you observe about the tree?
2. What colors do you see?
3. What are the parts of the tree? [roots, trunk, leaves, branches, twigs, etc.]
4. What are some changes you noticed?
5. Do you think the tree will look the same if you look at it next week? ...next year? ...when you are grown up? Explain.
6. What changes in the tree were fast changes? What changes were slow changes?
7. What are you wondering now?

Spinning Through The Seasons

Topic
Seasonal change

Key Question
How do plants change throughout the seasons?

Learning Goals
Students will:
- identify changes in the appearances of plants during the different seasons,
- arrange a set of pictures so that the seasons go in the correct order, and
- create a season change wheel for a plant.

Guiding Documents
Project 2061 Benchmarks
- *Change is something that happens to many things.*
- *Things change in some ways and stay the same in some ways.*
- *Some events in nature have a repeating pattern. The weather changes some from day to day, but things such as temperature and rain (or snow) tend to be high, low, or medium in the same months every year.*

NRC Standard
- *Ask a question about objects, organisms, and events in the environment.*

Science
Earth science
  seasons
Life science
  seasonal changes

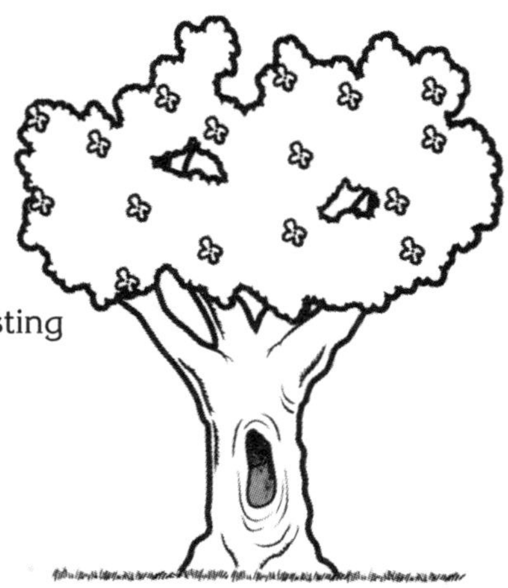

Integrated Processes
Observing
Comparing and contrasting
Identifying

Materials
For the class:
  scissors
  glue sticks
  tape
  card stock
  colored pencils, optional
  seasonal tree photos, optional

For each student:
  craft sticks
  paper fasteners
  season change wheels, one set
  plant cards, one set

Background Information
As the seasons change, so do the appearances of plants, trees, bushes, and so on. In areas with lots of deciduous trees, the changes in the seasons are particularly noticeable. An awareness of how plants and trees change in the different seasons will provide the young learner with a foundation for later learning about the reasons for those changes. This activity provides students with an opportunity to apply their observations to put a series of pictures in order, giving them a way to see the seasonal change as a repeating cycle.

Management
1. Each student should have his or her own seasonal change wheel. Each wheel requires a copy of each circle page, a craft stick, and a paper fastener.
2. Copy both season change wheel pages onto card stock for durability. You will need to use the tip of a sharpened pencil to punch through the dark circle in the center of each wheel before giving them to students.
3. Depending on the abilities of your students, you may wish to cut out the wheels and the pictures before beginning this activity.
4. Plan to display the sets of plant cards using a projection device.
5. Digital photographs of trees in each season are provided. If you have a computer with a projection system, you can use these photos in *Procedure 5.*

Procedure
1. Ask several students to name a plant, tree, flower, bush, etc., that they like. Ask each one why they like that particular plant.
2. Select a student whose response had to do with a characteristic that is seasonal. (e.g., I like our cherry tree because it's pretty and I like to eat cherries.) Ask the class in what season(s) this person probably likes the plant best. [spring for the flowers, and summer for the cherries]
3. If no students gave a usable response, give an appropriate example of your own and have the class

identify the season that corresponds to your description. (e.g., I like maple trees because of how beautiful they are when their leaves change color.)

4. Have students brainstorm some generalizations about how plants look in each season. Try to get a variety of plants—deciduous trees, evergreen trees, flowers, bushes, etc. Discuss how some plants change more than others do from season to season. (oak tree versus pine tree)
5. If you have pictures of plants up in your room, have students identify which season is pictured and why they think that. If you have a computer with a projection system, use the provided digital photographs.
6. Tell students that they are going to put some pictures in order so that they go in a sequence from winter to spring to summer to fall.
7. Distribute one set of plant cards to each student. If necessary, have students cut out the cards. Allow time for all students to get their cards in order.
8. Display one set of plant cards using a projection device. Invite one student who had that plant set to come up and identify the card he or she placed first in the sequence and explain the choice. See if the other students who had that same plant agree with the choice, and if not, come to a consensus.
9. Continue to invite additional students up, one at a time, to continue the sequence until all eight cards are in the correct order going from winter to fall. Repeat this process with the other plant card set.
10. Distribute the season change wheels, craft sticks, paper fasteners, tape, and glue sticks to students. Go through the procedure for assembling a season change wheel.
11. Provide assistance as students glue down their pictures and assemble their wheels.
12. Discuss how putting the pictures on a wheel shows that the seasons move in a repeating pattern. After winter, it's spring; after spring it's summer, and so on.

Connecting Learning

1. How do plants change in the different seasons?
2. What would be different about an apple tree in the fall and in the spring?
3. What kinds of plants change a lot from season to season? [flowers, deciduous trees, etc.] What kinds of plants don't change very much from season to season? [evergreen trees, shrubs, etc.]
4. What kinds of things do you look for to determine which season it is when looking at a picture of a plant? [the leaves, flowers, fruit, etc.]
5. How did you decide in what order to put your cards? Did everyone agree on the order of the cards? Why or why not?
6. What does the season change wheel tell you about how plants change? [It's a repeating pattern. Every spring the apple trees get new leaves, and every fall, they lose their leaves.]
7. What are you wondering now?

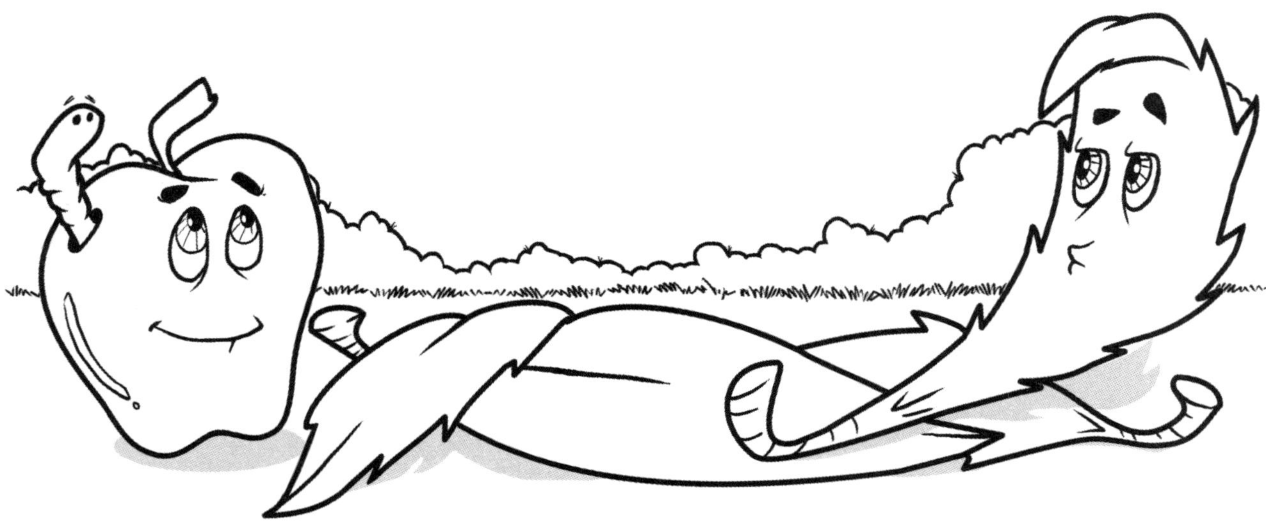

Spinning Through The Seasons

Spinning Through The Seasons

Spinning Through The Seasons

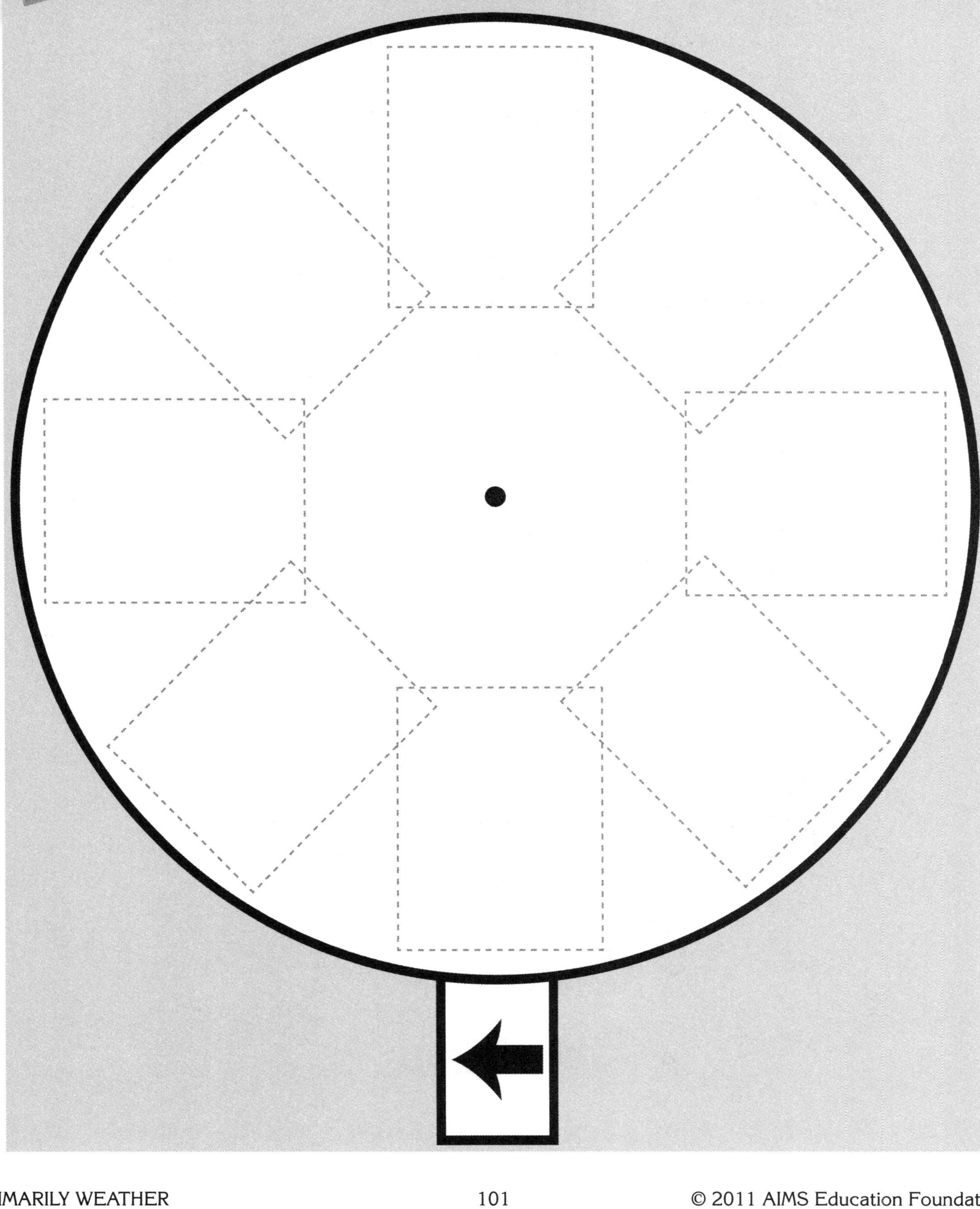

Spinning Through The Seasons

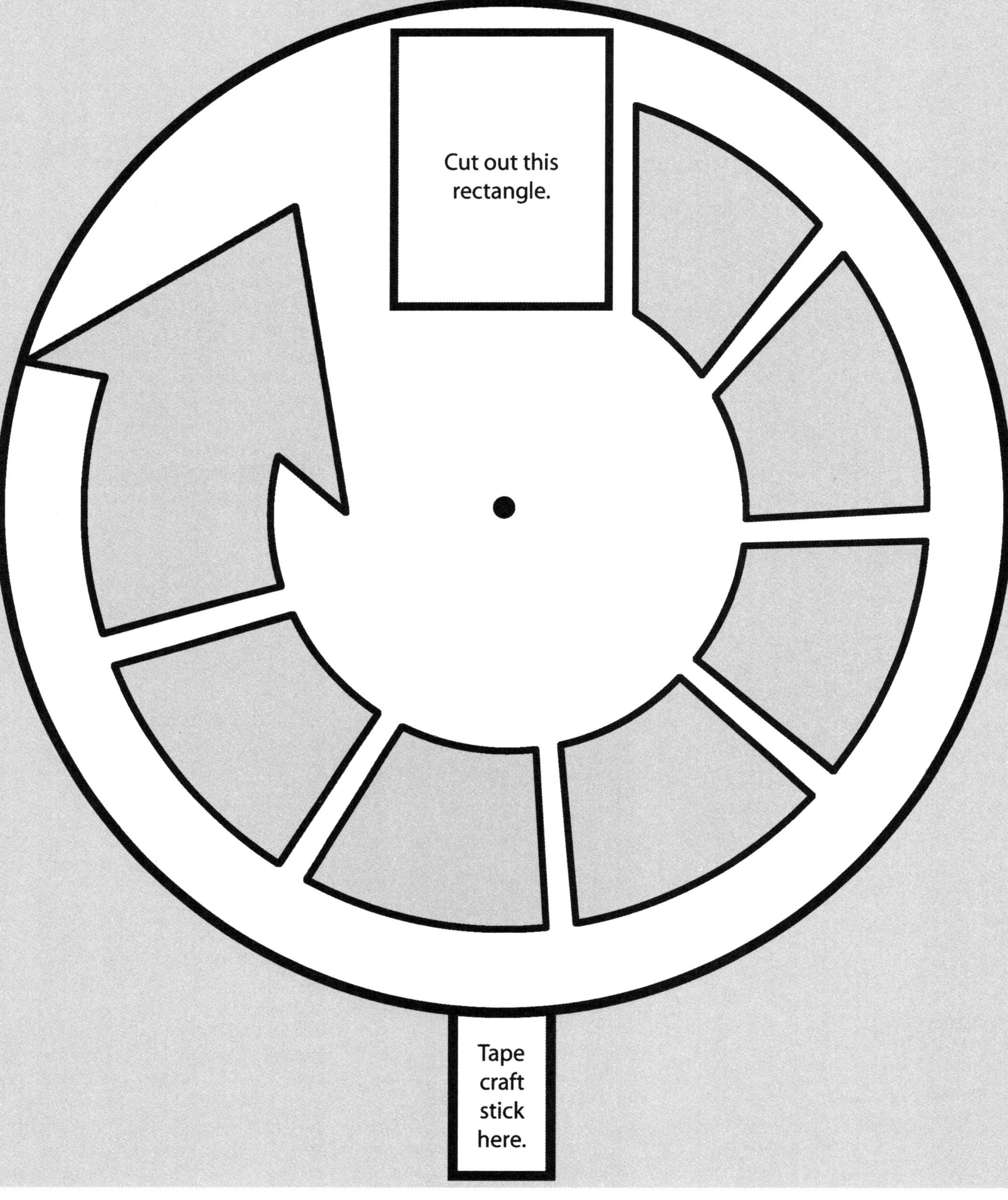

Season Cycles

Topic
Seasonal adjustments

Key Question
How do the seasons change how animals act?

Learning Goals
Students will:
- recognize that changes in seasons change animal behavior, and
- identify what season it is by how animals behave.

Guiding Documents
Project 2061 Benchmarks
- *Some events in nature have a repeating pattern. The weather changes some from day to day, but things such as temperature and rain (or snow) tend to be high, low, or medium in the same months every year.*
- *Plants and animals have features that help them live in different environments.*

NRC Standard
- *The behavior of individual organisms is influenced by internal cues (such as hunger) and by external cues (such as a change in the environment).Humans and other organisms have senses that help them detect internal and external cues.*

Science
Earth science
  seasons
Life science
  animals
    adaptations

Integrated Processes
Observing
Comparing and contrasting

Materials
Student pages
Scissors
Tape or glue

Background Information
Animals change their activities with the seasons for health and survival. Birds migrate to find available sources of food. Squirrels eat (summer) and store food (fall) while it is available. This stored food allows them to survive through winter and spring until new food sources are available. Bears eat large quantities of food in the summer months and develop a large surplus of fat by fall. As they hibernate during the winter, the fat provides their energy. By spring, when they begin to forage for food, they have lost their fat. While an animal's behavior is controlled by instinct that responds to seasonal changes, humans also modify their behavior with the seasons. These conscious changes made by humans can be seen most easily in change of clothing, adapting to the changing temperature of the seasons.

The pictures of the four seasons show the animals doing the actions found in the chart at the end of this teacher's text.

Management
1. Copy a set of season pictures and one page of children for each group of four students. These may be laminated for repeated use.
2. Copy an assessment page for each student.

Procedure
1. Distribute the four season pictures to each group of four students.
2. Have each group put the pictures in order starting with the present season.
3. Focus the students' attention on the first picture. Ask the students to identify the animals in the picture and what they are doing. Have the students suggest what about the season might encourage the animals to exhibit the behavior.
4. In sequence, have the students compare how each of the animals' behaviors has changed, and discuss how the change of the season might have encouraged the changed behavior.
5. Have each group place the eight pictures of children on the appropriate seasonal pictures. Then have the class discuss where they put each of the children and what clues allowed them to correctly place them.
6. Starting with the present season and going in order, have the students discuss how the children have changed for the season and how that change is like the animals' change.

7. To assess student learning, distribute the page with the four pictures of the bears and four pictures of children. Have the students cut apart the eight pictures. Have students make pairs of the bear and child that represent the same season. After making the four pairs, have them put them in sequence and attach them to a paper, one on top of the other. Under each pair have the students write the season that is represented by the pair. (A word list of seasons might be put on the board if spelling is a difficulty.) When students have completed their pages, have them discuss how they know their pairs are correct.

Connecting Learning

1. What are the ducks (squirrels, bears) doing in this picture?
2. Why during this season might the ducks (squirrels, bears) be doing what they are doing?
3. How has what the ducks (squirrels, bears) are doing in this picture changed from what they were doing during the season before this one?
4. What changed in the season to make the ducks (squirrels, bears) change what they are doing?
5. How do you know which children go with each season?
6. How do the children change between seasons?
7. How is the change in the children like the change in the animals this season?
8. What are you wondering now?

	Spring	Summer	Fall	Winter
Ducks	Flying and returning (migrating) to the pond as warm weather is returning.	Swimming in pond and feeding on available food sources.	Leaving and flying away (migrating) to find warmer weather.	None in the area. Have migrated to warmer locations.
Squirrels	Active outside, but new food sources are limited.	Eating new food sources that are available.	Gathering nuts as a food source for the long winter.	Finding shelter from winter weather and eating stored nuts.
Bears	Coming out of hibernation with little fat.	Eating berries that are abundant to develop excess fat.	Seeking a warm location to hibernate after storing food as fat.	Hibernating in cave to conserve energy through winter.
Humans	Dressing in short sleeves as the weather begins to warm while they examine new plant growth, flowers.	Playing in the water during the warm weather. During the hot weather they need to stay cool.	Wearing sweaters as weather cools.	Wearing coats, snowsuits, and boots to stay warm while playing in the winter.

Season Cycles

Season Cycles

Assessment

Temperature Told - Hot or Cold

Topic
Temperature

Key Question
What does a thermometer tell us about temperature?

Learning Goal
Students will learn how thermometers show temperature by building a model and by using an immersion thermometer.

Guiding Documents
Project 2061 Benchmark
- *A model of something is different from the real thing but can be used to learn something about the real thing.*

NRC Standard
- *Objects have many observable properties, including size, weight, shape, color, temperature, and the ability to react with other substances. Those properties can be measured using tools, such as rulers, balances, and thermometers.*

Science
Physical science
  temperature

Integrated Processes
Observing
Communicating
Comparing and contrasting

Materials
For each student:
thermometer headband (see *Management 4)*
thermometer model
strip of red construction paper (see *Management 3)*
tape
scissors
student page

For each pair of students:
immersion thermometer
one transparent cup of ice
one transparent cup of very warm water

Background Information
Heat and temperature are related, but they are not the same thing. Most people are familiar with the idea of temperature. The temperature of our bodies is important for our health, recipes tell us the temperature of the air in the oven for baking, the weather report gives us the temperature of the air outside. Our experiences tell us that the warmer something feels, the higher the temperature is likely to be.

A thermometer is an instrument used for measuring temperature, usually by expansion or contraction of mercury or alcohol in a tube and bulb. CAUTION: Use only alcohol thermometers when working with young students.

Heat energy flows from warmer to cooler. If two objects are put into contact, and we wait until all changes stop, the objects will be at the same temperature. If we put the end of a thermometer into something hot, heat energy will flow from the substance to the thermometer and it will come to the same temperature as the substance. No matter where we put the thermometer, the energy will flow in or out and make the thermometer the same temperature as the substance it touches.

Management
1. Immersion thermometers are available from AIMS (item number 1976). These thermometers have a V-shape structure and are more durable for use with young students.
2. It is best to do this activity with partners so thermometers can be shared. Each student should make a model of the thermometer and make his or her own predictions in the activity.
3. Prepare the red construction paper strips for the thermometer models ahead of time by cutting 3" x 11" strips.
4. You may choose to prepare the thermometer headbands ahead of time, or students can do the cutting and assembly. To make the thermometer headbands, cut out the thermometer and cut apart the headband strips. Tape the strips together to make one long strip, and tape the thermometer to the center. Tape the ends of the strip together around each child's head so the headband is snug.
5. Caution students not to rapidly move their thermometers from a substance of one temperature to something with a radically different temperature. They should wait a few minutes before immersing the thermometer into the new substance.
6. This activity begins with the review of the terms *hot* and *cold*. The students will discuss various items that are either hot or cold and place the pictures under the correct headings.

7. The students then will dramatize the movement of the red column in a thermometer so that they can observe and relate to the changes in models and actual thermometers. It is important to go back and forth from the model to the actual thermometer as described in *Procedure* as well as using the dramatization. The dramatization is repeated during the activity to reinforce the movement of the actual column of liquid in a thermometer. These diverse experiences are a way of getting understanding for the various learning styles in the classroom.

Procedure

1. Review the terms *hot* and *cold*. Discuss the pictures at the bottom of the student page. Ask the students how they know whether each item is hot or cold. Have them relate their experiences with each item. Direct the students to cut out the pictures and glue them in the correct box.
2. Hold up a thermometer or the enlarged thermometer and ask the students how this tool helps them learn about things. Point to the red column on the thermometer and ask them what that column tells them, what happens to it, and why. If available, read *Temperature and You* (see *Curriculum Correlation).*
3. Distribute the thermometer headbands or have students make them (see *Management 4).* Once they are done, have them put the headbands on and act out changes in temperature. For example: Direct students to squat down and tell them it is very cold, so cold it may snow. Then describe the sun coming up and the air warming. Ask them what should happen to the thermometer's red column (their headbands). [The red column moves up.] Continue with other scenarios of changes in temperature as the students act out the changes by raising and lowering their bodies.
4. Demonstrate how to make the model of the student thermometer and distribute the necessary materials to students.
5. Once students have made their models, lead them in a discussion that will connect the dramatization they did with the model of a thermometer they just made. Focus their attention on what happens to the red column in each case.
6. Hold up an immersion thermometer and show the students how to properly use it. Talk about the glass portion and how it can break if misused. Tell the students they are going to be using the thermometer as a tool to learn about the temperature of things. Caution them not to move the thermometer from hot things to cold things too quickly.
7. Distribute to each student pair a transparent plastic cup with ice in it. Ask the students to predict what will happen to the red column in the immersion thermometer when they put it into the ice. Have the students show their predictions first with their body position (upright or squatting) and then with their model thermometers. Once they have predicted with their models (adjusted red column), say "everybody show" at which time all students should hold up their thermometers so you can scan the group.
8. Distribute the immersion thermometers (one to each pair of students), and then direct them to carefully immerse them into the ice and observe what happens. (They should watch it for at least three minutes to allow it to stabilize.) After observing the changes in the red column, have them adjust the red column in their models to match what has happened to the real thermometer. Again, say "everybody show" and scan the group to assess their understanding of the changes in the thermometer.
9. Have students remove the thermometers from the ice and allow time for the thermometers to stabilize to room temperature. Repeat the procedure using very warm water, predicting with their bodies and the models, using the real thermometers, and then adjusting the models.
10. Discuss what they learned and how they can use this skill of reading a thermometer.

Connecting Learning

1. What kinds of things are hot? What kinds of things are cold?
2. What happened to the red line on the thermometer when you put it in the cup of ice? [It went down.] What does this tell you? [The ice is colder than the air temperature.]
3. What happened to the red line on the thermometer when you put it in the cup of warm water? [It went up.] What does this tell you? [The water is warmer than the air temperature.]
4. Why do we use thermometers? [To tell us how hot or cold things are—to tell us their temperatures.] When have you seen or used thermometers before? [Outside to measure air temperature, when I am sick to measure my body temperature, etc.]
5. What do you think would happen to the red liquid in the thermometer if you moved it from inside the classroom to outside the classroom right now? Explain. [If the temperature outside is warmer, the red liquid will go up. If the temperature outside is colder, it will go down. If it is the same, it will do nothing.]

6. Which do you think would be colder—a cup of water from the refrigerator or a snow cone? What could you do to find out? [Use a thermometer and see which one makes the line of red liquid go lower.]
7. What are you wondering now?

Extensions

1. Use the immersion thermometer in other things like lukewarm water, sand, or soil.
2. Apply this understanding to the use of an indoor/outdoor thermometer.

Curriculum Correlation

1. Read: Maestro, Betsy and Giulio. *Temperature and You.* Lodestar Books. New York. 1990.
2. Describe the changes in temperature when writing in your journal.

Home Links

1. Have students take their model thermometers home and explain them to their families.
2. Encourage students to find thermometers around their house and start using them to find out about the changes in the temperature.

Temperature Told – Hot or Cold

Hot	Cold

 ice cream	 stove	 cocoa	 ice water	 ice cream bar
 soup	 freezer	 candle flame	 snow	 fire

TEMPERATURE TOLD– HOT OR COLD HEADBAND

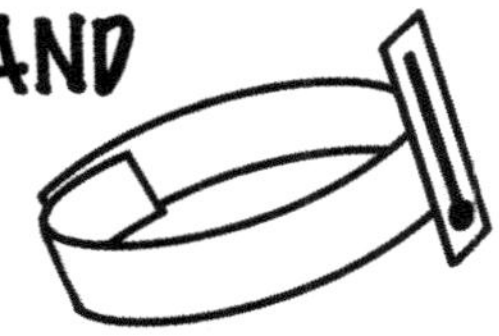

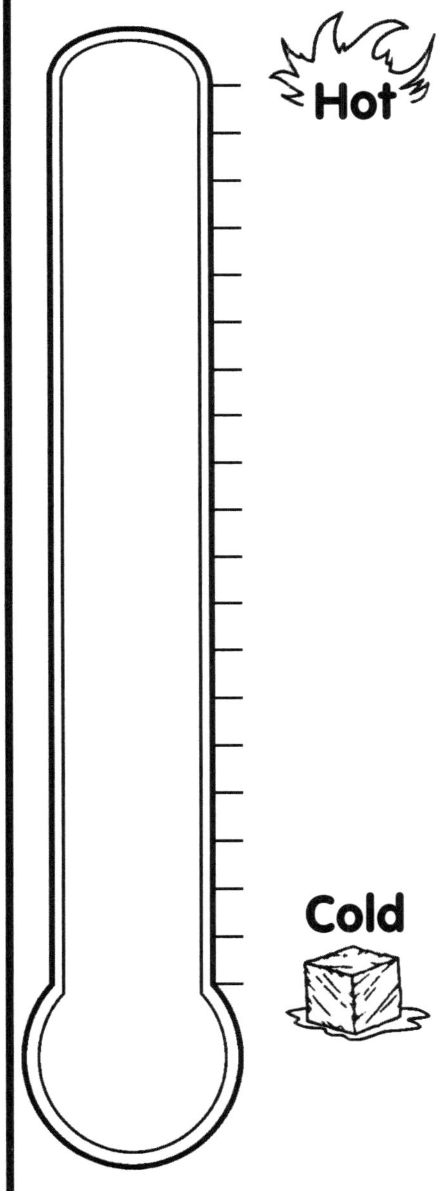

Name

Temperature Told– Hot or Cold

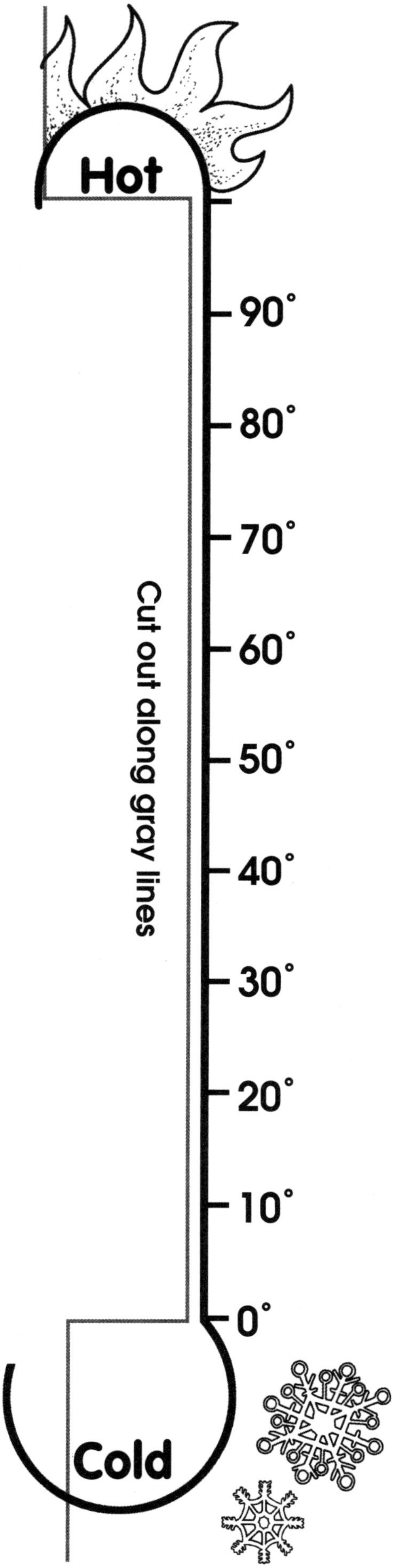

Fold along this line second

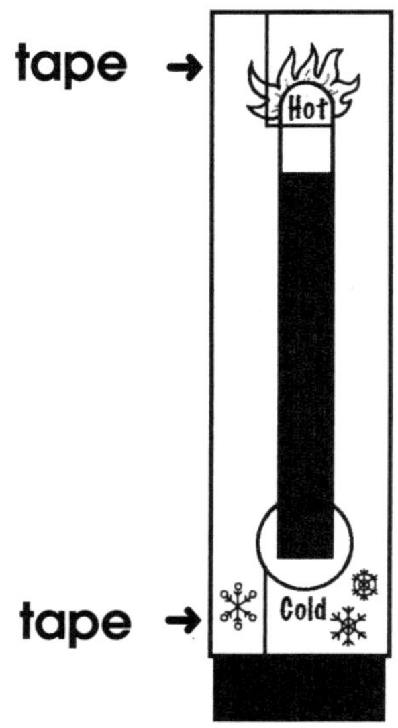

Fold along this line first

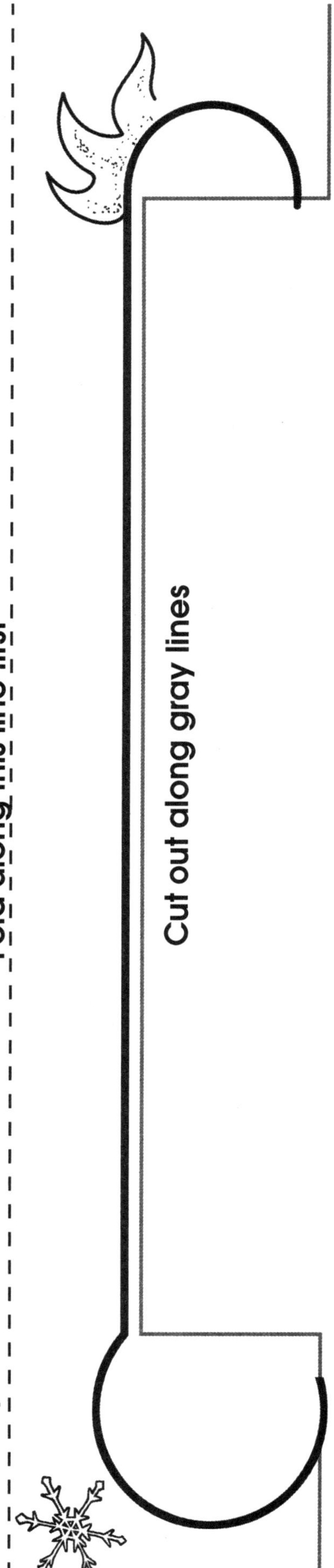

Directions

1. Cut out along gray lines.
2. Fold along dashed lines.
3. Tape as indicated to hold paper together.
4. Cut 3" x 11" strip of red paper.
5. Insert paper strip into the middle of the thermometer.

The Temperature Today

Topic
Weather: Temperature

Key Question
What is the temperature today?

Learning Goals
Students will:
- collect local temperature data from the Internet twice daily for one week,
- look for patterns in the data, and
- keep track of temperature data throughout the year.

Guiding Documents
Project 2061 Benchmark
- *Some events in nature have a repeating pattern. The weather changes some from day to day, but things such as temperature and rain (or snow) tend to be high, low, or medium in the same months every year.*

NRC Standards
- *Weather changes from day to day and over the seasons. Weather can be described by measurable quantities, such as temperature, wind direction and speed, and precipitation.*
- *Employ simple equipment and tools to gather data and extend the senses.*

Science
Earth science
    weather
        temperature

Integrated Processes
Observing
Collecting and recording data
Comparing and contrasting
Generalizing

Materials
Computers with Internet access
Crayons or colored pencils
Student page

Background Information
Weather changes from day to day and with the seasons. One of the ways to keep track of weather data is by looking at the temperature. In this activity, students will use the Internet to collect real-time temperature data for their location twice a day. This data will be recorded, allowing students to recognize any patterns and trends that may be present. For example, it is (usually) warmer in the afternoon than it is in the morning. If there is a warming trend while you collect data, the afternoon temperatures will rise as the week progresses. A storm system may move in, causing a rapid, drastic change in temperature. While young students are not able to understand the reasons for these changes, they can recognize the patterns that the data present. As they collect data throughout the year, they will begin to recognize seasonal changes in the temperature and become better at connecting the temperature to how hot or cold it feels outside.

Management
1. This activity begins with collecting data over a period of one week, but it is assumed that this data collection will continue over the course of the year in conjunction with a weather calendar.
2. You will need to begin this activity on a Monday so that five consecutive days of data can be collected. Students will need to collect data twice a day—once in the morning and once in the afternoon. Space these two collection times as far apart as your school day will allow. The data should be collected at about the same time each day.
3. Before beginning the activity, go to the *Weather Report for Kids* website (http://kidsweatherreport.com/) and enter your location's zip code. Bookmark the resulting page. Do this on each computer that students will be using.
4. Plan to display the student page using a projection device while introducing the activity.

Procedure
Part One: Introduction
1. Ask students if they have ever watched the weather on the evening news or heard the weather reported on the radio. Have them share some of the things that are mentioned in the weather report. [temperature, humidity, wind, cloud cover, precipitation, etc.]

2. Talk about the temperature portion of the weather report. Discuss what the temperature tells you. [how hot (or cold) it will be that day] To verify that students recognize that higher numbers correspond to warmer temperatures, ask several questions comparing two temperatures. For example, Which day is warmer, one where the high temperature is 85°, or one where the high temperature is 55°? Which day is colder, one where the high temperature is 32° or one where the high temperature is 47°?
3. Ask the students if they have heard what the temperature is supposed to be today. Ask them where they got this information. [newspaper, TV, radio, parents, etc.] If no one knows, ask them how they could find out.
4. Explain that students will be keeping track of the temperature twice a day for one week. They will get their information from the Internet.
5. Distribute the student page and display it using a projection device. Invite students to study the thermometers. Have them notice that the marks on the scale go up by five degrees at a time. There is a longer line every 10 degrees, and numbers are written every 20 degrees.
6. Write a temperature in the first space and color the thermometer to correspond to the temperature. Discuss with students how you determined how to color the thermometer.
7. Write a temperature in the second space and ask students how you should color the thermometer if this is the temperature. Invite a student to come to the overhead and color the thermometer accordingly. Ask if everyone agrees with how it was colored, and make any changes that might be necessary. Repeat this process until you are confident that students understand how to record approximate temperatures on the thermometer.

Part Two: Data Collection

1. Invite students to gather around a computer that has the weather website on the screen. Ask them to identify the current temperature. [The current temperature is at the top in parentheses. The forecasted low and high are underneath the picture of the children.]
2. Instruct students to record the current temperature in the space for Monday morning and color the thermometer accordingly. Allow them to go outside for a minute to feel what this temperature is like.
3. Repeat this process in the afternoon, and then every morning and afternoon for the remainder of the week. For these subsequent data collection times, students can use the computers in pairs or groups and go to the website to collect the temperature data.
4. Discuss how the temperature is related to the general weather conditions and what patterns or trends students see in the data they have collected.
5. Continue to gather temperature data throughout the year as part of a weather calendar experience and make comparisons of temperature from season to season.

Connecting Learning

1. What does the temperature portion of a weather forecast tell us? [how hot or cold it will be that day]
2. What was the temperature in the morning on Monday? How did this compare to the temperature in the afternoon?
3. Did you notice any patterns in how the morning temperatures compare to the afternoon temperatures? (Usually, it is warmer in the afternoon than it is in the morning, but this is not always the case.)
4. Was it hotter on Tuesday afternoon or Wednesday afternoon? How do you know?
5. Were there any patterns in the afternoon temperatures? Describe them.
6. Which morning was the coldest? How do you know?
7. Were there any patterns in the morning temperatures? Describe them.
8. Were there any days when we had the same temperatures in either the morning or the afternoon?
9. Put the days of the week in order from coldest to hottest afternoon temperatures. How did you decide on this order?
10. What did you notice about the weather and how that related to temperature? (For example, clear sunny days are warmer than cloudy days.)
11. How does the temperature change with the seasons? Would you expect warmer or colder temperatures in the fall/winter/spring/summer?
12. If I told you that it is 85° outside, would you want long sleeves or short sleeves? Would you want a jacket? Would you want shorts or long pants? What if it was 45°? …100°?
13. How would you describe each of those temperatures (warm, cold, hot, etc.)? Does everyone agree on how a certain temperature feels? Why or why not? (Perceptions of hot and cold are subjective. They are also affected by other factors such as wind, cloud cover, and humidity.)
14. What are you wondering now?

Extensions

1. Compare the actual high and low temperatures for each day to the temperatures recorded. How much colder/hotter did it get?
2. Collect data on another aspect of weather such as wind speed or precipitation.

Internet Connections

The website recommended for this activity is *Weather Report for Kids* because it has a simple, kid-friendly interface and only gives students temperature data. The other websites are additional places where more complete weather information can be found.

Weather Report for Kids
http://kidsweatherreport.com/

CNN Weather
http://weather.cnn.com/

National Weather Service
http://www.weather.gov/

Weather Underground
http://www.wunderground.com/

Weather Channel
http://www.weather.com/

	Monday	Tuesday	Wednesday	Thursday	Friday
Morning	______ °F	______ °F	______ °F	______ °F	______ °F
Afternoon	______ °F	______ °F	______ °F	______ °F	______ °F

Air Temperature

Topic
Using a thermometer

Key Question
How does air temperature affect a thermometer?

Learning Goal
Students will use a thermometer to measure air temperature.

Guiding Documents
Project 2061 Benchmarks
- *Tools such as thermometers, magnifiers, rulers, or balances often give more information about things than can be obtained by just observing things without their help.*
- *Some events in nature have a repeating pattern. The weather changes some from day to day, but things such as temperature and rain (or snow) tend to be high, low, or medium in the same months every year.*

NRC Standards
- *Weather changes from day to day and over the seasons. Weather can be described by measurable quantities, such as temperature, wind direction and speed, and precipitation.*
- *Simple instruments, such as magnifiers, thermometers, and rulers, provide more information than scientists obtain using only their senses.*

Math
Measurement
 temperature

Science
Earth science
 weather
  temperature

Integrated Processes
Observing
Predicting
Comparing and contrasting
Collecting and recording data
Communicating

Materials
Immersion thermometers
Small containers for water
Warm water
Cool water
Masking tape, optional
Permanent colored pens, optional

Background Information
Temperature tells us how hot or cold something is. Most people are familiar with the idea of temperature. The temperature of our bodies is important to our health; the temperature of the air is given in the weather report. Our experience tells us that the warmer something feels, the higher the temperature is likely to be.

A thermometer is an instrument whose size, shape, or some other feature changes when its temperature changes so that it can be used to measure temperature. The most common thermometers are those that have an expanding column of mercury or colored alcohol. (Only alcohol thermometers should be used in elementary school.)

Management
1. Large Styrofoam cups or liter boxes work well for the water containers.
2. Make sure the warm water will not be uncomfortable to the touch.
3. If students have not had prior experience with thermometers, you will need to spend time during the lesson talking about the scale of the thermometers.
4. It is best to use thermometers with only one scale. The student page illustrates both Fahrenheit and Celsius scales. You may want to cover one of the scales before copying the page to prevent confusion.
5. Thermometers (item number 1976) are available from AIMS.

Procedure
Observing the thermometer
1. Give each group a thermometer. Have the students examine the thermometer and discuss their observations.
2. Explain to the students that the thermometer has colored alcohol in the bulb that expands (takes

up more room) when it gets warm or hot and the liquid goes up the tube. Encourage them to offer ways they can get the alcohol to go up the tube.

3. Ask them what they think will happen if they put their thumbs on the bulbs of the thermometers. Allow time for them to test their predictions. Ask them what they know about the temperature of their fingers. [If the alcohol rose in the thermometer, their thumbs were warmer than the surrounding air.] Have them remove their thumbs from the thermometers' bulbs and, after a short time, ask them what has happened to the column of liquid and why it happened.
4. Provide the students with two containers of water, one warm and one cool. Tell them to stick one hand in each container and note the difference. Have them predict what will happen when they place their thermometers in the two different containers of water.
5. Direct them to place the bulb ends of their thermometers in the containers of water. Urge them to observe what happens to the red liquid.
6. When the red liquid stops rising, ask the students which number the red liquid is next to.
7. Encourage them to compare their results with those of other groups and to record their observations.

Using the thermometer for air temperature

1. Distribute the student page.
2. Have the students take their thermometers outside and record the temperature of the air in the sun and in the shade.
3. Ask the students to explain the differences in the temperatures.
4. Encourage them to find other areas that would have big differences in temperatures, for example, the blacktop and a grassy area. Have the students explain why they think one area will be hotter than the other.

Connecting Learning

1. What do you like to eat or drink that is hot? What is your favorite cold food or drink? How does each of these foods become hot or cold?
2. Do you have thermometers at home? What are they used for? [air conditioning, heater or furnace, body temperature]
3. When something is warm, what happens to the thermometer? When something is cold, what happens to it?
4. When you were using your thermometer, what did you notice about the red liquid in the tube? [It goes up when it is hot, and it goes down when it is cold.]
5. When you put your thumb on the bulb of the thermometer, what happened to the red liquid? What temperature were you measuring? [my thumb's]
6. When you put the thermometer in the containers of water, what temperature were you measuring? [the water's]
7. When you measured the temperature of the air outside, where did you find the coolest temperature? ...the warmest?
8. How can you tell if something is hot or cold without actually touching it yourself?
9. What are you wondering now?

Extension

Challenge the students to find the hottest and the coldest spot on the school ground. Let them explain why they chose certain areas to investigate.

Curriculum Correlation

Maestro, Betsy and Giulio. *Temperature and You.* Lodestar Books. New York. 1990.

Air Temperature

Measure Air Temperature

Color in the red to match your thermometer.

Where: ____________

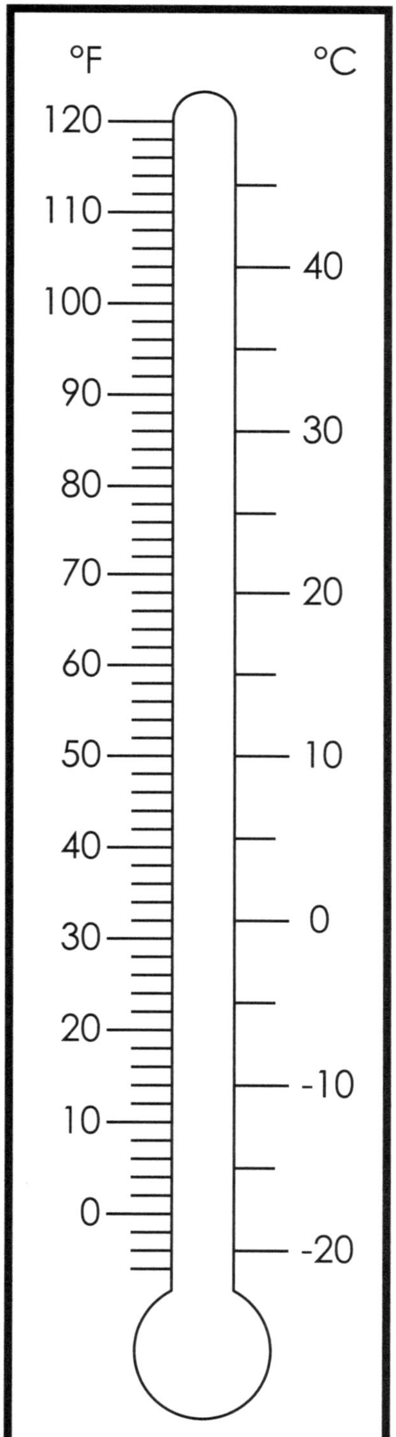

Where: ____________

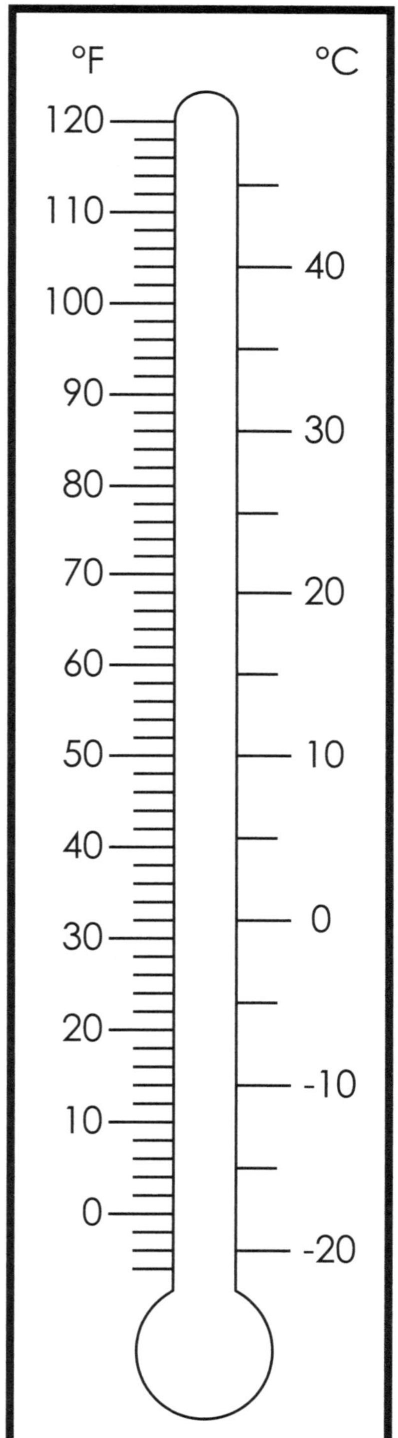

Where: ____________

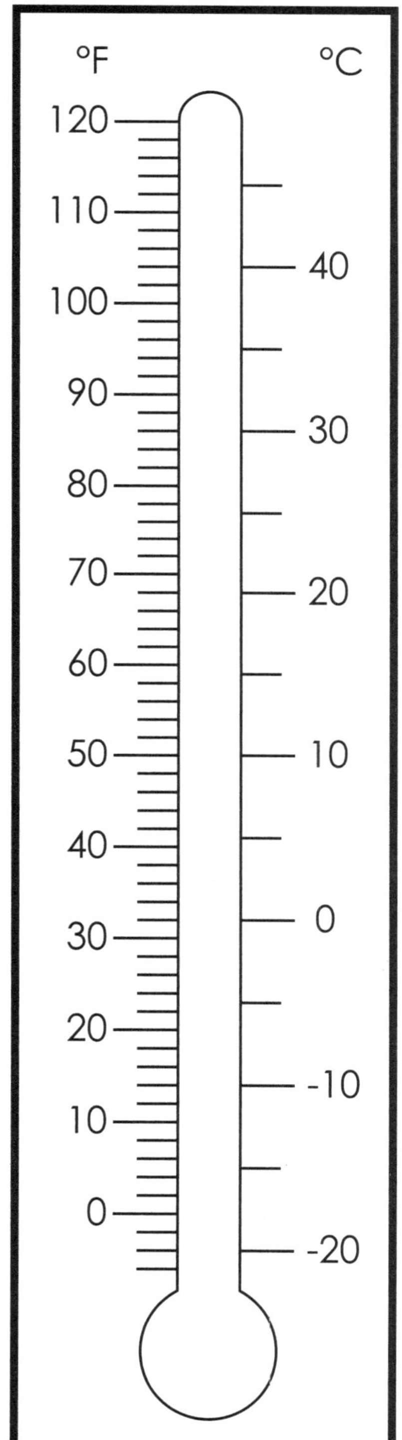

What did you find out?

Just Right

Tune: Polly Wolly Doodle

Optional Verses

When I play hard and need a drink
Cold milk sure hits the spot;
If it's too cold, though, it might freeze,
Or spoil if it's too hot. (chorus)

In winter if it gets too hot,
There's rain instead of snow;
But sometimes I must stay inside
While icy cold winds blow. (chorus)

In summer if it gets too hot
I just don't want to play;
But sometimes it might be so cold
that I can't swim all day. (chorus)

Hand motions may be used to accompany this song. Have the children pretend that they are thermometer. Whenever something too hot is mentioned, they should reach up high; when they sing about something too cold, they should reach down low. At the just right parts, they should make sweeping motions with their hands at their sides. Add other familiar motions as you wish.

Wind Detectives

Topic
Wind

Key Question
What can we observe about wind and how it moves things?

Learning Goals
Students will:
- observe and describe evidence of wind,
- observe that objects are moved by the wind and that different objects move in different ways,
- draw pictures of things that are moved by the wind and sort them based on a variety of rules, and
- create a bar graph to display the pictures.

Guiding Documents
Project 2061 Benchmarks
- *People can often learn about things around them by just observing those things carefully, but sometimes they can learn more by doing something to the things and noting what happens.*
- *Things move in many different ways, such as straight, zigzag, round and round, back and forth, and fast and slow.*
- *Things move, or can be made to move, along straight, curved, circular, back-and-forth, and jagged paths.*

*Common Core Standards for Math**
- *Represent and interpret data. (1.MD/2.MD/3.MD)*

Math
Data collection
- graphing
  - bar graph
- sorting

Science
Earth science
- weather
  - wind

Physical science
- force and motion

Integrated Processes
Observing
Comparing and contrasting
Collecting and recording data
Communicating

Materials
Part One:
- *Gilberto and the Wind* (see *Curriculum Correlation)*
- various objects to observe in the wind (see *Management 3)*

Part Two:
- crayons or colored pencils
- scissors
- glue
- bulletin board paper or chart paper (see *Management 4)*
- student pages
- magazines, optional

Background Information
The wind provides us with visible evidence of the invisible air around us. Wind is a result of the unequal heating of the Earth's surface that creates areas of high and low atmosheric pressure. Wind exerts a force on all objects it encounters. If this force is large enough to overcome the other forces acting on an object (friction and gravity), the object will move. This activity provides students with the opportunity to make some initial observations about the wind and how it moves different objects in different ways. Students should be encouraged to observe the changes in the wind day by day and over time. Emphasis should be placed on students making careful observations.

Management
1. This activity should be done on a windy or breezy day.
2. This activity works best as a total group activity followed by small group and individual explorations.
3. Prior to doing this activity, gather a collection of objects that will be moved easily by the wind such as pieces of paper, Styrofoam cups, plastic bags, leaves, feathers, tissue, and so on. Also collect a few items that will not move easily in the wind

(rocks, staplers, textbooks, etc.) and a few that may move depending on how they are positioned (round pencils, small paperback books, etc.).

4. You will need a piece of chart paper or bulletin board paper large enough to use as a whole-class bar graph on which to organize and display students' drawings from *Part Two.*

Procedure

Part One

1. Have students discuss what they know about wind. Encourage the students to look out the window and decide if the wind is blowing. Have them discuss the evidence that the wind is blowing.
2. Read the story *Gilberto and the Wind* (see *Curriculum Correlation)* and ask the students if they have ever played in the wind. Ask how they can tell that the wind is blowing in the story and what things it moves.
3. Brainstorm some objects that the wind can move and record them on a chart or the board.
4. Take the students on a walk outside to observe evidence of the wind on the playground. [moving trees, blowing swings, flag flapping, dust devils, etc.] Discuss the different ways in which the wind can be felt and seen.
5. Return to the classroom and distribute to the students as many of the items from the list as possible. Supplement the items brainstormed with other items suggested in *Management 3* so that each child has one item. At least two or three children should have objects that will not move as easily in the wind, and two or three children should have objects that may or may not move depending on how they are placed (e.g., round pencils, paperback books, empty margarine tubs, etc.).
6. Go to a blacktop or other relatively smooth surface that is exposed to the wind. Have each student place his or her object on the ground so that the class can observe how the wind affects it. (Be sure to have students pick up any objects that do not belong in the natural environment once they have been observed.)
7. Have students with objects that were not moved by the wind experiment with different ways to place the objects that might cause them to be moved. For example, a round pencil would not roll when placed parallel to the wind, but might roll when placed perpendicular to the wind.

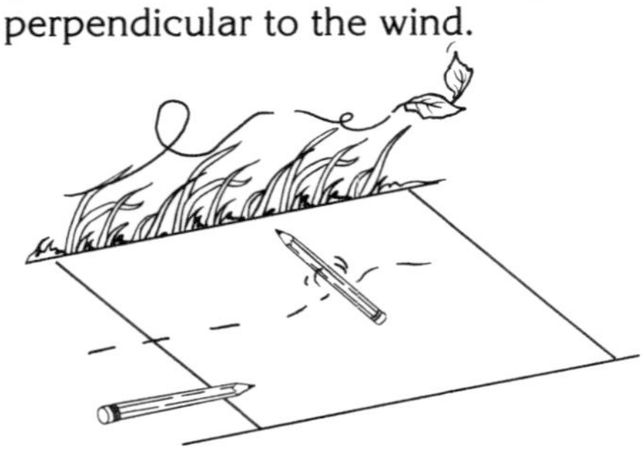

8. Help students verbalize that the wind moves some things more easily than others. Help them recognize that even the way an object is placed can change whether or not it will be moved by the wind. Have them use descriptive language to compare and contrast the different ways in which the different objects moved.

Part Two

1. Take the students back inside and distribute the first student page and the crayons or colored pencils.
2. Direct the students to draw a picture in each box to record something the wind can move. (One of the four pictures should be of something that was not observed in *Part One.*) Pictures can be cut from magazines and pasted into the boxes.
3. When students have finished, direct them to cut their papers into the four sections. Have them work in small groups to sort the pictures into two or more groups. Rules for the sorts can be given by you as the teacher or come from students, but they should all have to do with the objects in relation to the wind and how they moved. Examples of sorts could include: things that the wind moved all the time, things that the wind moved some of the time, and things that the wind never moved; things that moved quickly in the wind and things that moved slowly in the wind; things that moved along the ground and things that moved in the air, etc.
4. When the groups have sorted their pictures, invite them to share their sorting strategies. Continue to share ideas and encourage as many sorting options as possible.
5. As a class, decide on one sorting scheme to use for all the pictures. Make a bar graph on a sheet of chart paper or bulletin board paper in which to display the pictures. Give the graph an appropriate title and label each section.

6. Place the graph on a low table or on the floor. Have the groups come up one at a time and place their drawings in the appropriate sections on the graph. Once all groups have sorted their drawings, have the class gather around the graph and decide if they all agree with the placement of the pictures. If there are any disagreements, have the students who placed the pictures explain why they made the choices they did.

7. When there is consensus about the organization of the graph, glue down each picture so that the graph can be displayed on the wall.
8. Distribute the second student page and have students either write or draw in the space provided at least two things that the wind can do. Encourage students to draw upon their experiences from the activity when completing this page.

Connecting Learning

Part One

1. What evidence of the wind did we see on the playground? [tree branches moving, hair blowing, flags waving, dust in the air, etc.]
2. What were some of the objects that moved in the wind?
3. What were some of the objects that did not move in the wind? Why didn't they move? [too heavy, wind wasn't strong enough, etc.]
4. What were some objects that only moved if they were placed a certain way? [pencil, cup, etc.] How did they have to be placed so that they would move? [facing the wind, on their sides, etc.]
5. How were the objects that moved in the wind different from those that did not move? [lighter, less stable, smaller, etc.]
6. Describe the different ways our objects moved in the wind. [some moved along the ground, some blew up into the air, some flapped or fluttered without moving very much, some moved quickly, some moved slowly, etc.]
7. What could you do to keep the wind from moving an object? When would you want/need to do that?
8. Why would it be important to know if the wind is blowing? Who are the people who need to know?

Part Two

1. What pictures did you choose to draw? Why?
2. What were some of the ways your group sorted your pictures? Why did you choose those sorts?
3. What does our graph tell us about the objects that the wind can move?
4. What are some of the things the wind can do?

Extensions

1. Have students take some of the objects from *Part One* outside on a windy day and observe what they do when they are not placed on the ground. How many of them still move when they are held in the hand? Give students ribbons, crepe paper, or similar objects that will flutter when held and have them compare and contrast the characteristics between the two sets of objects.
2. Have students brainstorm and experiment with objects that may or may not move in the wind depending on the state they are in. For example, a closed umbrella would probably not move in the wind, while an open umbrella likely would. Other objects could include inflatable balls, balloons, or cardboard boxes (broken down vs. assembled).
3. Help the students make their own wind scale. For example:

   No Wind—the flag hangs limply, the trees are still

   Light wind—the wind felt on face, the leaves are barely moving

   Moderate wind—light flag extended, small branches move, pieces of paper and trash are blown around

   Strong wind—large branches move, flag is flapping vigorously

   Dangerous wind—difficult to walk against wind, small branches broken off trees

Curriculum Correlation

Ets, Marie Hall. *Gilberto and the Wind.* Penguin Puffin Books. New York. 1978.
Gilberto has fun playing in the wind.

Simon, Seymour. *Let's Try It Out in the Air.* Simon & Schuster. New York. 2001.
This book for young learners encourages children to think about the world around them and engages them in fun and simple activities.

Wind Detectives

Wind Detectives

The wind can...

The Wind Blows

Topic
Wind

Key Question
What can we observe about the wind?

Learning Goal
Students will observe and measure how fast the wind is blowing.

Guiding Documents
Project 2061 Benchmark
- *Tools such as thermometers, magnifiers, rulers, or balances often give more information about things than can be obtained by just observing things without their help.*

NRC Standard
- *Weather changes from day to day and over the seasons. Weather can be described by measurable quantities, such as temperature, wind direction and speed, and precipitation.*

*Common Core Standards for Math**
- *Measure lengths indirectly and by iterating length units. (1.MD)*
- *Measure and estimate lengths in standard units. (2.MD)*

Math
Measurement
Ordering

Science
Earth science
  weather
    wind

Integrated Processes
Observing
Comparing and contrasting
Collecting and recording data
Classifying

Materials
Scissors
Tissues
Pencils with erasers
Pushpins
Tape
Crayons
Rulers
Lightweight ribbon or crepe paper streamer (60 cm per student)
Small pieces of paper, Styrofoam, bits of material, small pieces of wood and plastic
Measuring tools (see *Management 3)*
Student pages

Background Information
Wind is moving air. When the air moves, you can feel it against your face. Wind can move large sailboats across the water. It can turn windmills to pump water. The wind can provide electrical power. Wind can blow so softly that it is not noticed, or it can blow so hard that it can damage buildings, trees, and ships. Wind can cause erosion by blowing away topsoil or by wearing away rock with sharp grains of dust.

The air moves because it has been warmed by the sun. Wind is caused by the uneven heating of the air by the sun. As the sun heats the air, it expands and rises. Air from cooler areas then rushes in to replace the heated air. Cold air is denser and has higher pressure; warm air is less dense and has lower pressure. Air flows from areas of high pressure to areas of low pressure.

We can't see air or wind, but we can see things that the wind is moving: particles of dirt or sand, leaves on the trees, the clouds scurrying across the sky, sailboats moving across the lakes, and the flag boldly waving. Students will not actually be measuring the speed of the wind (distance/time), instead they will be observing the movement of a ribbon in the wind and establishing their own scale for how fast the wind is blowing.

Management

1. Beforehand, cut lightweight ribbon or crepe paper streamers into 60-cm lengths. These will be taped on the ends of rulers to make wind measuring tools.
2. Urge students not to leave any litter from their investigations outdoors.

3. Cut and tape colored paper in meter lengths for measuring tools. Use a variety of colors so meter tapes placed end to end are of different colors. This will help students begin to build a concept of meter.

Procedure

Exploring Wind—Part One

1. On a windy day encourage the students to look out the window, or if they have just come in from outside, ask them, "Do you think it is windy outside today? Why do you think that?" Guide them to explain the objects they see being blown by the wind. Encourage them to explain what their opinion of a "windy" day is.
2. Discuss with the students what they know about wind. Can they see it? ...hear it? ...taste it? ...feel it? ...smell it?
3. Take the students outdoors. Give them a tissue and have them release it and run to keep up with it.

Exploring Wind—Part Two

1. Have pieces of paper, small pieces of a Styrofoam cup, a feather, a leaf, and small pieces of wood or plastic and ask the students to guess which ones will be moved the farthest by the wind. Allow time for the students to order the objects according to the distance they think they will be moved by the wind. On a windy day, have them test the objects. Ask them to determine how they could make "fair tests" and establish some class rules. An example might be to hold the object over their head, let it go, mark by sticking a pencil in the ground where it lands, and measure the distance.
2. Urge the students to perform three trials. Ask them to explain how they decided which object was carried the farthest. Have them record their results on the first student page.
3. Take the students outdoors and have them run in different directions. Ask them how the wind feels against their bodies. Ask them if they can move more easily with the wind or against the wind. Encourage them to explain their answer.
4. Have them describe the sounds they hear when the wind blows.

Wind Speed—Part Three

1. Invite the students to brainstorm words to describe the action of the wind: gentle, light, breezy, mild, moderate, blustery, whipping, strong, harsh, and dangerous.
2. Ask the students if they can figure out how fast the wind is blowing. [look at the school flag, watch the trees and leaves blowing, or a loose piece of paper being blown] Help the students make their idea of a wind scale. For example:
   No wind—the flag hangs limply, the trees are still
   Light wind—the wind felt on face, the leaves are barely moving
   Moderate wind—light flag extended, small branches move, pieces of paper and trash are blown around
   Strong wind—large branches move, singing heard in wires
   Dangerous wind—difficult to walk against wind, small branches broken off trees
3. Have the students make their own wind measurer by taping a 60-cm length of ribbon or crepe paper to the end of a ruler.
4. While indoors, have the whole class stand in a circle holding their wind measurers up above their heads. Guide everyone to observe that if they stand still, the ribbon or crepe paper streamer hangs down.
5. Tell the students to walk in a circle so they create their own air movement or wind. Guide the students to observe what happens to the wind measurers when they move faster or more slowly. Ask them what they would have to do to make their streamers stay still or stand straight out.
6. Take the students outdoors during different wind conditions and have them observe the action of the wind measurers as they stand still.
7. Discuss how certain types of winds can be helpful or harmful to us. Be sure not to arouse fears of tornadoes, hurricanes, and other extreme conditions.
8. Distribute the second student page and the materials and assist students in making pinwheels.
9. Let the students test their pinwheels indoors by blowing on the wheels.
10. Take the pinwheels outdoors and let the students test the spin. Have them see if the wind is as strong as their lungs. Urge them to see if they can use the pinwheel to determine wind direction. Ask them if they think they can make some guidelines to know how fast the wind is blowing by using their pinwheels. Allow time for them to test their ideas.

Connecting Learning

1. How do you know when there is wind?
2. What comes to your mind when you hear, "Today is a windy day?"
3. What do you feel when the wind is blowing on your face?
4. Do you always have air around you? How do you know?
5. Describe how the trees move in the wind.
6. Is it correct to say that we see the wind blowing? What do we see? [things blown by the wind]
7. Were you able to keep up with your tissue? Did it blow in the air or along the ground?
8. What things did you find on the school grounds that were blown by the wind?
9. Describe the action of your wind measurer.
10. What other ways can you tell how fast the wind is blowing?
11. What are you wondering now?

Extensions

1. Display a wide variety of pictures that show wind or moving air. Have the students take two of the pictures and compare them as to how hard they think the wind is blowing. You may also wish to have the students decide which pictures show useful wind (as perhaps sailing), and which shows a harmful wind (a destructive storm).
2. Take the students for a walk on a windy day. Encourage them to look for signs of the wind. Ask questions such as these: Are the trees moving in the wind? How do they move? Do the leaves move? What else on the tree is moving? Does the wind in the trees make a sound? What do you think is making that sound?

Curriculum Correlation

Language Arts

1. Read *Gilberto and the Wind* by Marie Hall Ets. Penguin Puffin Books. New York. 1978. Discuss with the class how much fun Gilberto had playing in the wind. Suggest that the students might want to make their own book describing their own adventures with the wind.
2. Read the Winnie the Pooh adventure *The Blustery Day* by A.A. Milne.
3. On a windy day, ask the students to observe what the wind is blowing: grasses, leaves, paper, students' clothes. Have the students suggest words that describe the movements. Encourage the students to move to show how they would explain the words.

Art

Drop paint onto slick paper. Blow air through a straw to make wild, windy paint pictures. Blow the paint so that it moves in different directions to create a painting. The paper can be used to make gift wrap.

Home Link

Have students take home their wind measurers and explain how to use them to their parents. Suggest that parents and students keep a record for a week of how fast the wind blows each day.

The Wind Blows

You need: leaf, feather, piece of paper, piece of foam cup

Which one will the wind carry farthest?

Do this:

1. Hold the object.

2. Let go.

3. Mark and measure.

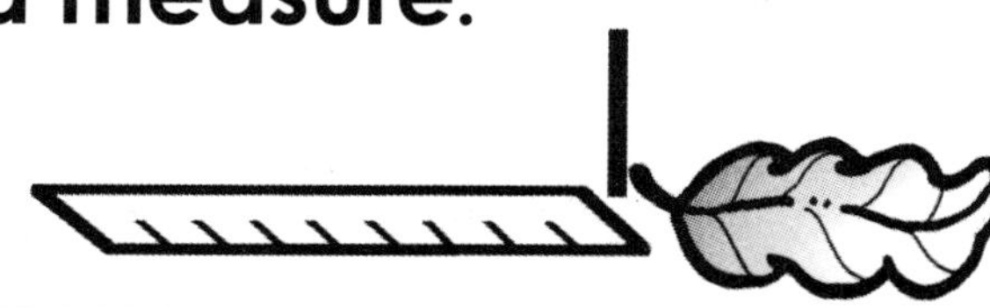

Object	Measure			Order
	Drop 1	Drop 2	Drop 3	
1.				
2.				
3.				
4.				

The Wind Blows

You need: pencil with eraser, pushpin, crayons, tape, scissors

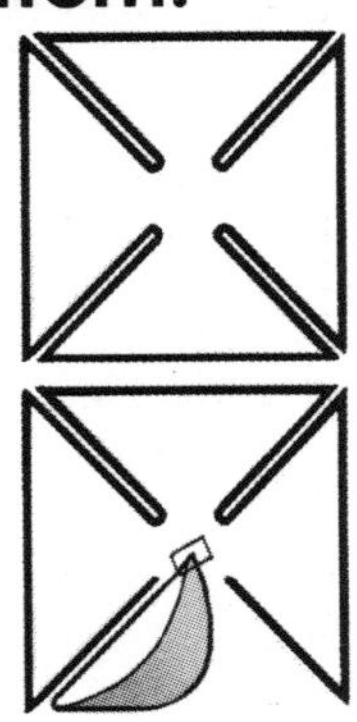

1. Decorate the pinwheel pattern.
2. Cut out the pattern.
3. Cut on the inside lines.
4. Bring the corners with the stars to the center. Tape.
5. Put the pushpin through the middle of the pinwheel into the eraser.
6. Test the wind outside with your pinwheel.

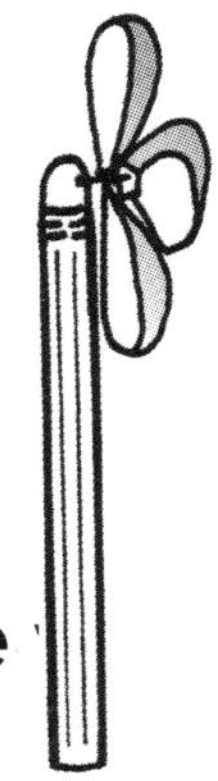

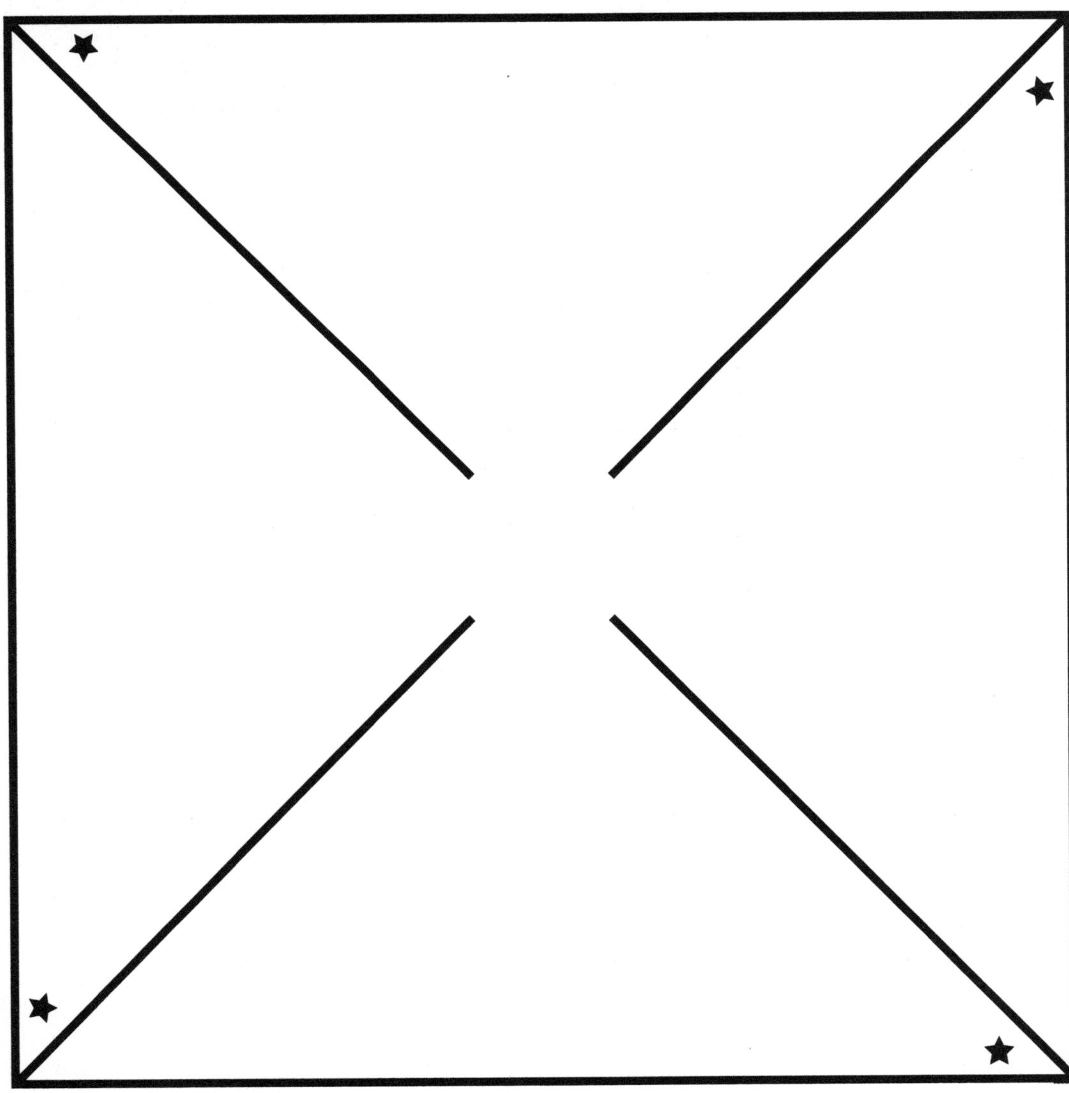

Which Way?

Topic
Wind direction

Key Question
How can you tell from which direction the wind is blowing?

Learning Goal
Students will make a wind vane to determine wind direction.

Guiding Documents
Project 2061 Benchmarks

- *People can often learn about things around them by just observing those things carefully, but sometimes they can learn more by doing something to the things and noting what happens.*
- *Tools are used to do things better or more easily and to do some things that could not otherwise be done at all. In technology, tools are used to observe, measure, and make things.*
- *A model of something is different from the real thing but can be used to learn something about the real thing.*
- *Some events in nature have a repeating pattern. The weather changes some from day to day, but things such as temperature and rain (or snow) tend to be high, low, or medium in the same months every year.*

NRC Standard

- *Weather changes from day to day and over the seasons. Weather can be described by measurable quantities, such as temperature, wind direction and speed, and precipitation.*

Science
Earth science
  weather
    wind direction

Integrated Processes
Observing
Classifying
Predicting
Applying

Materials
Scissors
Glue
Drinking straws
Straight pins
Paper clips
Pencils with erasers
Tagboard
Tape

Background Information

We usually think of wind as air that moves along Earth's surfaces. As the sun heats the air, it expands and rises. Cooler, denser air moves in to take its place. This movement of air is wind.

Wind direction is the direction from which the wind comes. We can usually tell from which direction the wind is blowing by watching things as they move. A wind vane is used to indicate the direction from which the wind is blowing. It does this by pointing into the wind.

Winds are named for the direction from which they come. A north wind is blowing from the north to the south. A wind vane points into the wind, so if it points to the north, the wind is a north wind.

A typical wind vane consists of an arrow on a pivot. The wind blows the larger end of the arrow away from the source of the wind, causing the smaller, pointed, end to point in the direction from which the wind is blowing. For many years, people have referred to wind vanes as weather vanes probably because they know the readings are closely related to weather changes.

Management

1. Wind vanes can be assembled by small groups working directly with the teacher.
2. Use extreme caution working with the straight pins.

Procedure

1. On a windy day, instruct the students to look out the window and to name as many things as they can that tell them that the wind is blowing. [leaves moving, flag flying, movement of the low clouds, dust and dirt blowing, papers and trash moving on the ground, etc.] Ask them if they can point in the direction from which the wind is coming.
2. Take the students outside and ask them to determine the direction from which the wind is blowing. If they need assistance, suggest that they wet a finger and hold it up. Ask them if one side of their finger feels cooler than the other. Inform them that this is the direction from which the wind is blowing. If they have their wind measurers from the activity *The Wind Blows*, invite them to hold them up. Allow the students time to determine that the wind is blowing from the direction opposite the tail

of the ribbon. Ask the students to turn so they face the wind.

3. Play the game *Simon Says* using wind directions. For example, Simon says, turn so the wind blows in your face. Turn so the wind blows on your back. Simon says, turn so the wind hits your right side, etc.
4. Explain to the students that the direction of the wind is shown by a wind vane. Describe a wind vane and ask if anyone has seen one (perhaps on top of a barn or house).
5. Take the students inside. Inform them that they are going to follow the directions on the sheet *Make a Wind Vane* so they can make their own wind vanes.
6. After the students have finished constructing their wind vanes, test them with an electric fan.
7. Take the students and their wind vanes outside to see how they work.
8. After they have seen a wind vane work, ask them if they can think of people who need to know the direction of the wind. [pilots, sailors, parachutists, hunters, air traffic controllers, farmers, weather forecasters]

Connecting Learning

1. Why is it important to know what direction the wind is blowing?
2. As you observed the wind, could you tell without using a wind vane what direction the wind was blowing? Explain.
3. How does a wind vane work?
4. Where do you think wind vanes would be used? Why would they be important at an airport?
5. What things do you like to do on windy days? What things do you like to do on calm days? Are any of them the same as what you do on windy days?
6. What are you wondering now?

Extensions

1. Another type of instrument used to indicate wind direction is a wind sock. These are often used at small airports. The wind sock fills with air and indicates the direction of the wind and the speed of the wind. The more horizontal the sock, the stronger the wind is.
   A wind sock is like a sock with a hole cut in the toe; the wind blows through the sock. There are many colorful wind socks on the market today. Hang one outside the classroom so that students can observe the speed of the wind as measured by the wind sock.
2. Brainstorm with the students all the things that fly in the air or use wind and air currents to move. [birds, insects, kites, balloons, clouds, seeds, gliders, etc.] Explain to the students that we use the wind in many ways: to turn windmills and pinwheels, to dry clothes, to fly kites, etc.
3. Record all the ideas and group them into natural or human-made categories. Have the students bring pictures from home of these things or draw pictures and make a *Using the Wind* bulletin board.

Curriculum Correlation

Language Arts

1. Show a picture of people in a very strong wind. Have the students write a story about the picture.
2. Have the students write a creative story about wind. An example of a writing prompt: *One day I was sitting all by myself when the wind came along and lifted me high over the trees. I felt lighter than air…* Finish the story and tell what you saw, how you felt, and how you got back.

Literature

Bauer, Caroline Feller. *Windy Day*. J. B. Lippincott. New York. 1988.

Home Link

Tell the students to take their wind vanes home and teach their families about wind direction.

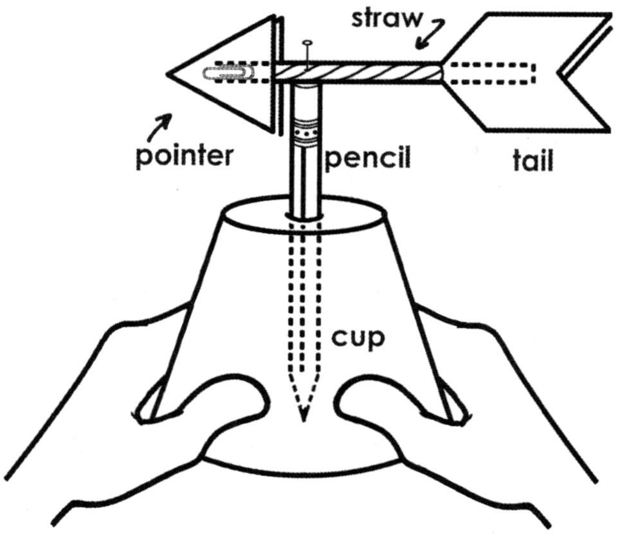

Which Way?

You need: straw, pin, pencil with eraser, paper clip, tagboard, cup

Do this:

1. Trace and cut out the pattern on tagboard or stiff paper.
   Make two tails and two pointers.

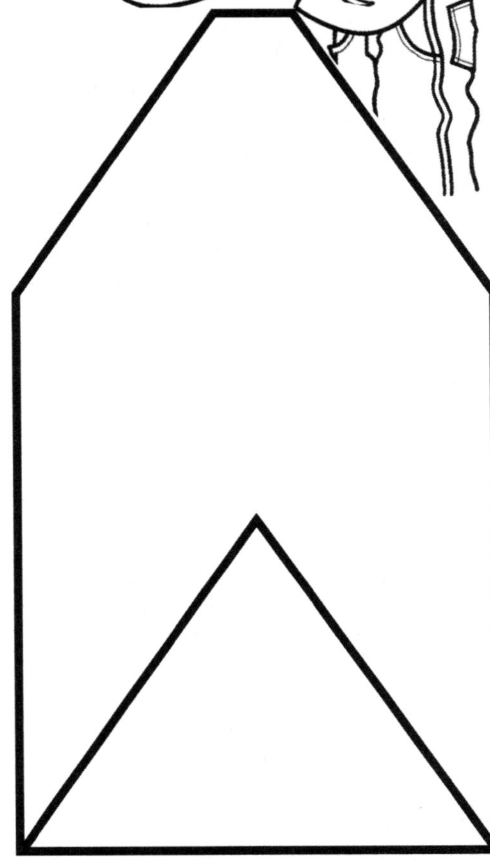

2. Put a paper clip on the end of a straw for weight.

3. Push a pin through the straw into a pencil eraser.

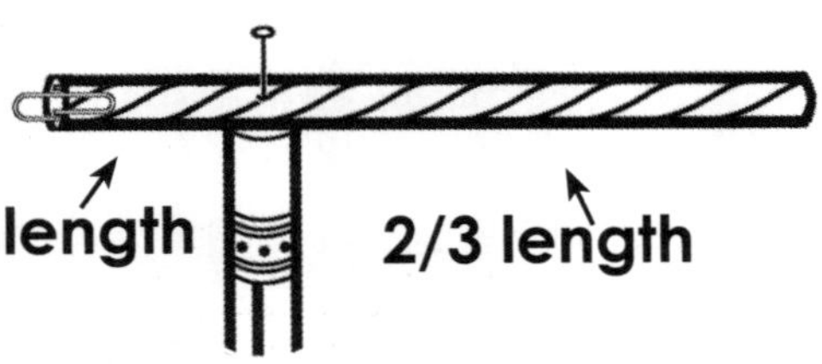

4. Glue the two pointers together on the paper clip end of the straw.
5. Glue the two tails together on the other end of the straw.
6. Push the pencil through the bottom of the cup and use as a holder.
7. Test to find wind direction.

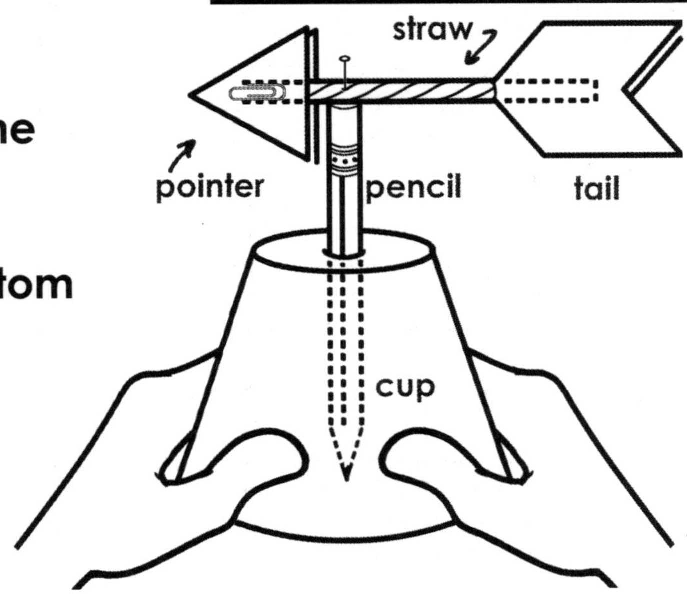

Cloudy Weather

Topic
Clouds

Key Question
What can you observe about clouds?

Learning Goal
Students will observe and classify clouds.

Guiding Documents
Project 2061 Benchmarks
- *Some events in nature have a repeating pattern. The weather changes some from day to day, but things such as temperature and rain (or snow) tend to be high, low, medium in the same months every year.*
- *The sun warms the land, air, and water.*

NRC Standards
- *Weather changes from day to day and over the seasons. Weather can be described by measurable quantities, such as temperature, wind direction and speed, and precipitation.*
- *The sun, moon, stars, clouds, birds, and airplanes all have properties, locations, and movements that can be observed and described.*

*Common Core Standards for Math**
- *Represent and interpret data. (1.MD/2.MD/3.MD)*

Math
Graphing

Science
Earth science
weather
clouds

Integrated Processes
Observing
Classifying
Predicting
Comparing and contrasting
Communicating
Applying

Materials
Cotton balls
Blue paper
Glue
Empty, clear two-liter bottle
Student pages

Background Information

Clouds are formed from water that has evaporated from rivers, ponds, oceans, and lakes. This water vapor expands and rises. As the water vapor expands, it cools. When the water vapor has cooled enough, it condenses and forms a cloud.

Clouds are masses of small water droplets that float in the air. There is always moisture in the air, but it forms into clouds only when the conditions are right. The tiny droplets of water need the right air temperature and something to cling to, such as dust, salt, or smoke particles. As the water forms into droplets, they seem to float in the sky, moved by the wind.

Each type of cloud carries a message, giving advance notice of coming changes in the weather. Clouds get their shape and form at certain heights because of the movement of air and the amount of water vapor and condensation particles in the air. In 1803, a British scientist used Latin names to identify cloud shape and height above ground: cirrus, stratus, and cumulus. Modern classification has added alto and nimbus. Most clouds can be identified with these names and their combinations.

The names that identify cloud shapes are *stratus* (spread out or layered), *cumulus* (heaped or piled up, lumpy), and *cirrus* (curly, wispy). The prefixes *alto* (middle level) and *cirro* (high level) indicate height above ground. *Nimbus* and the prefix *nimbo* refer to clouds that produce precipitation.

Stratus (spread out) are relatively low level, layered clouds that often cover the sky with a blanket of gray clouds. They may be found spread out, anywhere from the ground's surface up to a height of about one mile. On the ground, these clouds are called fog. Stratus clouds often indicate that rain is coming.

Middle-altitude clouds use the prefix *alto*. *Altostratus* clouds are located higher than stratus, but have a similar formation. *Altocumulus* (high heaps) are located at approximately the same height, but have the cumulus shape. They appear as small cloud patches arranged in rounded heaps of whitish masses.

The highest clouds are called *cirrus* (curl of hair). Cirrus clouds are the high, wispy white clouds made up of feathery ice crystals. They indicate a change in the weather. The prefix *cirro* is used to identify high level clouds of different shapes, such as *cirrocumulus* and *cirrostratus*.

Cumulus (heap or pile) are puffy, white clouds with flat bottoms that indicate fair weather. If they are building, it may turn stormy. *Cumulonimbus* (dark rain clouds) are clouds that can stretch up thousands of feet from their base and contain huge amounts of moisture. They are heaped-up piles of clouds. These kinds of clouds are often called thunderheads. Thunder, lightning, and rain often come from them.

Management

1. This activity should be done over a period of time so that students can observe various cloud formations.
2. Keeping a special calendar with illustrations of the clouds and other weather observations for each day may help students to see that clouds can be used as predictors of weather.
3. The illustration of the major types of clouds (*Cloud Types*) is included for your information. If desired, you can display the *Cloud Types* page using a projection device when students begin to talk about the classification of cloud shapes and height.
4. The focus of this lesson is NOT on learning the names of the different clouds. "Naming" should not be considered an indication of "knowing" in this lesson. Instead, the focus is one of observation. Students should observe that clouds have different shapes ("clumpy, straight, wispy") and appear to be at different heights. If the students desire to know the names, these should be considered incidental.
5. If the names of the clouds are not used, alter the graph to include only the illustrations of the different formations.

Procedure

1. Be certain that students have had the opportunity for focused observation of different types of skies: a clear day, a partly cloudy day, a cloudy day.
2. On a partly cloudy day, take the students out on the school grounds to sit down and watch the clouds. Tell them to use their imaginations and describe what shapes they can see. Urge them to continue their descriptions as the clouds change shape. Ask them why they think the clouds change shape. Ask them to share other observations.
3. Back in the classroom, have each student complete the sentence "A cloud looks like..." The answers can be recorded on the board or on a piece of butcher paper cut in the shape of a cloud.

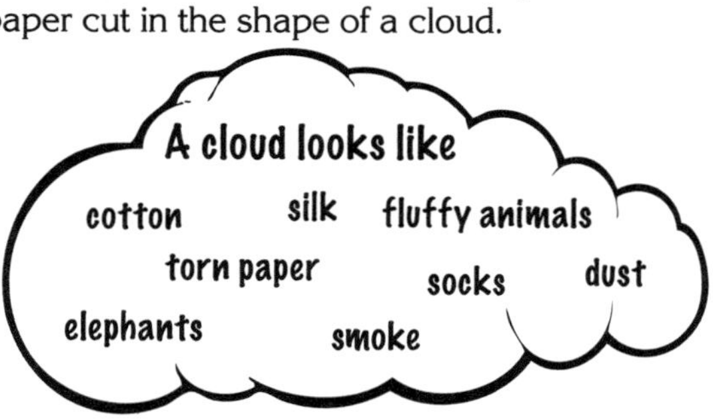

4. On another cloudy/partly cloudy day, have them follow up by completing the first student page.
5. Use the students' illustrations to sort clouds into groups such as *puffy clouds, straight layered clouds, feathery or wispy clouds.* Tell students that people who study clouds also group them by how high they are.
6. Ask the students what they think clouds are made of. After allowing time for discussion, ask the students if they think they have ever created a cloud. If necessary, suggest that they think about their breath on cold days. Allow them time to realize that they make clouds on cold days by breathing out warm, moist air into the cold outdoor air. As the air cools, it briefly forms a little "cloud." Ask them if they have ever seen a "cloud" at the grocery store when they open the freezer door.
7. Tell the students that they may be able to see a cloud in a bottle if they look closely. Inform them that you will make the cloud because it requires hot water and you don't want them to be burned. Point out that the hot water is like their warm moist breath. Pour about a cup of hot water into an empty, clear two-liter bottle. Shake the bottle vigorously to warm the inside of the bottle. Pour out most of the water, leaving only about one capful. Place the bottle in front of the students and screw the cap securely in place. Have the students look through the bottle at the light in order to see the wisps of vapor as it is produced. Squeeze the bottle, then relax your grip. A cloud should form. If not, repeat the steps but before putting on the cap, strike a match and throw it into the bottle. (The smoke provides particles for condensation.) Cap the bottle. Squeeze it and then relax it. Ask the students what they see happening in the bottle.Have them record their observations on the second student page.

8. Use blue poster paper to make a sky on a wall or bulletin board. Assign different cloud types to the students so that all are not the same. Distribute blue construction paper. Have them use cotton balls glued to construction paper to make the different kinds of clouds. (If some assistance is needed, suggest that they stretch out the cotton to show the wispy clouds (cirrus) with blue sky showing through, bunch up the cotton for the puffy clouds (cumulus), stretch the cotton for layered clouds (stratus). Create a sky scene, putting the stratus close to the ground and cirrus up high.

9. Suggest that students become cloud watchers to see what story clouds have to tell. Inform them that weather watchers often use clouds to predict what kind of weather is coming.
10. Have students keep track of the cloud types by graphing the clouds in the sky each day for a month.
11. Challenge them to try to predict the weather by observing the clouds.

Connecting Learning

1. Are there always clouds in the sky? Are there clouds today?
2. Describe the shape of the clouds. How high are they—high, medium, or low?
3. Have you ever been in a cloud that is on the ground? Describe it. How did it feel to walk through it?
4. Have you ever watched the clouds move? What makes them move?
5. What exciting shapes have you observed in the clouds?
6. When there are clouds in the sky, what happens to the sun? [The sun is still there but is sometimes blocked from our view.]
7. When is it brighter outside, when there are clouds or when there are no clouds? Why do you think this happens?
8. If you were asked to predict the weather for the next day just by looking at the clouds, how would you do it?
9. What are clouds made up of?
10. What are you wondering now?

Extension

Research folklore that is connected with the appearance of different kinds of clouds, such as "Red clouds in the morning sailors take warning. Red clouds at night, sailors delight."

Curriculum Correlation

Language Arts

1. Read the book, *It Looked Like Spilt Milk* by Charles G. Shaw. Harper Trophy. New York. 1947.
2. Discuss with the students what shapes they have seen in the clouds. Have the students write a cloud book. Each page is illustrated to show a shape in the clouds that the students saw.

CLOUDY Weather
CUMULONIMBUS
wind
thunderstorms
CIRRUS
ice crystal clouds
change in weather
CUMULUS
white, puffy
fair weather
CLOUD TYPES
STRATUS
low, gray
rain or snow
called fog on the ground

Cloudy Weather

CLOUDS

1. Draw a picture of the clouds you see.

2. Describe the clouds you see.

3. What do you think makes clouds?

CLOUDY Weather

CLOUD IN A BOTTLE

1. Observe the cloud inside the 2-liter bottle.

2. Draw what you see.

3. Tell how to make a cloud inside of a bottle in 3 steps.

First,

Second,

Third,

Cloudy Weather

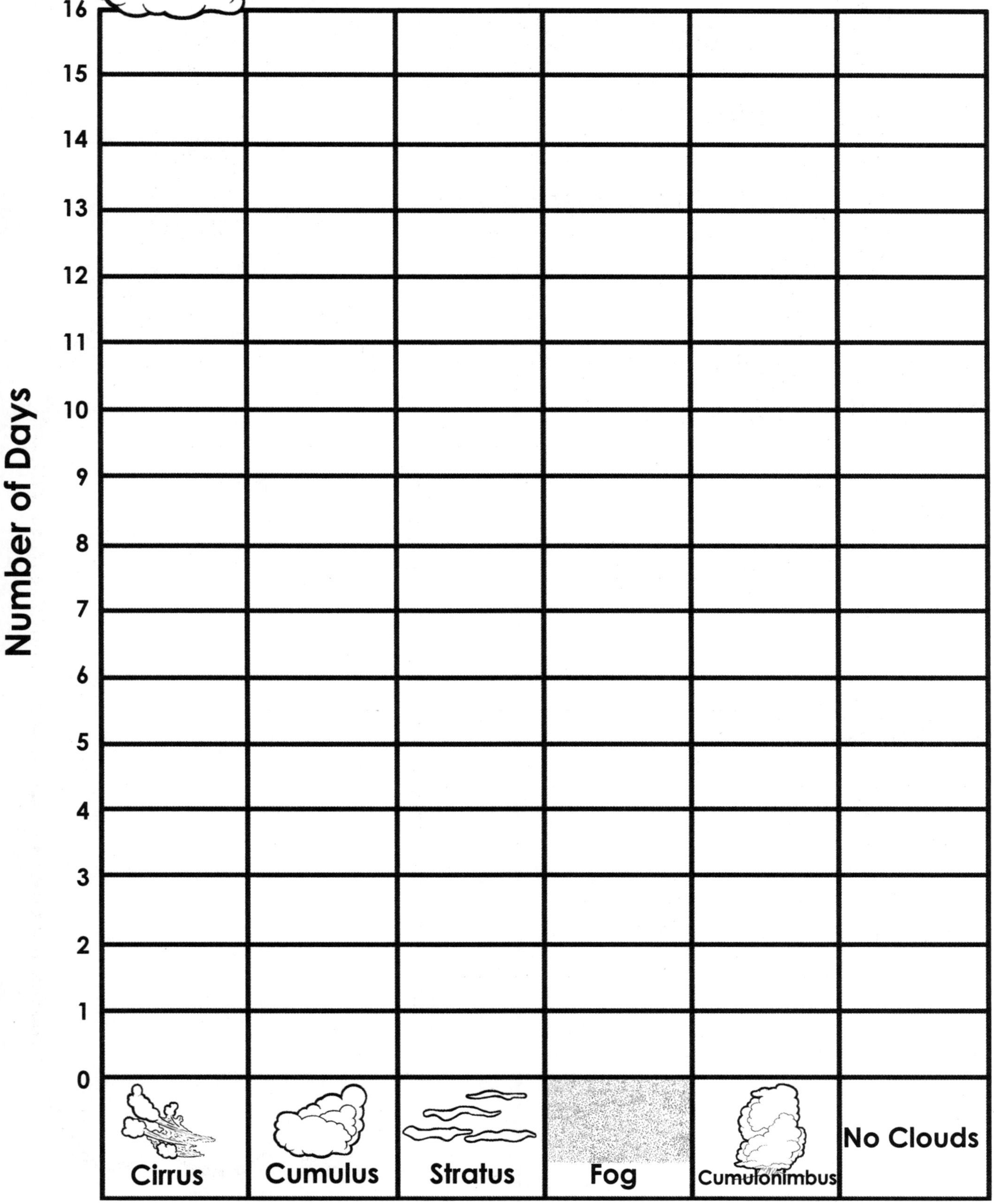

Station Precipitation

Topic
Precipitation

Key Question
How can we measure the amount of precipitation we received today?

Learning Goals
Students will:
- create a rain gauge,
- learn how to measure the amount of precipitation in the gauge, and
- measure the amount of precipitation received daily.

Guiding Documents
Project 2061 Benchmarks
- *Tools such as thermometers, magnifiers, rulers, or balances often give more information about things than can be obtained by just observing things without their help.*
- *Some events in nature have a repeating pattern. The weather changes some from day to day, but things such as temperature and rain (or snow) tend to be high, low, or medium in the same months every year.*

NRC Standards
- *Employ simple equipment and tools to gather data and extend the senses.*
- *Simple instruments, such as magnifiers, thermometers, and rulers, provide more information than scientists obtain using only their senses.*
- *Weather changes from day to day and over the seasons. Weather can be described by measurable quantities, such as temperature, wind direction and speed, and precipitation.*

*Common Core Standards for Math**
- *Measure and estimate lengths in standard units. (2.MD)*

Math
Measurement
Estimation

Science
Earth science
weather
precipitation

Integrated Processes
Observing
Collecting and recording data

Materials
Large bucket (see *Management 3)*
Dowel (see *Management 4)*
Ruler
Yardstick
Watering can

Background Information
Precipitation can take many forms, the most familiar to us being rain, snow, and hail. Measuring the amount of precipitation throughout the year not only provides valuable practice with measurement tools, it gives students a deeper understanding of the seasonal changes and patterns in the weather.

Management
1. This activity provides the background and practice for the precipitation recording section of the daily calendar. You will need to decide if you wish to collect data for a period of several weeks, a few months, or all year long.
2. Find a location on your school grounds where you will be able to place the rain gauge. The location should be out in the open away from any buildings or trees, but in a place where it will not be disturbed.
3. To make a rain gauge, you will need a large bucket or other similar container with a large opening and straight sides. Paint buckets or buckets from bulk laundry detergent work well. Most trash cans and typical cleaning buckets will not work because they have sides that are tapered, which distorts the accuracy of measurements.
4. To determine the amount of rain that has collected in the rain gauge, students will need dowels that will show a visible water mark. These dowels will be

placed in the bucket to see how deep the water is, and then laid next to a ruler to measure the number of inches of rain. Measurements only need to be made to the nearest quarter of an inch. Students use "less than" or "greater than" to be more specific.
5. To measure snow, you may want to simply put a yardstick in the snow until it touches the ground and take a reading. You will need to measure the depth in an area that has not drifted. If you want to measure the amount of precipitation from the snow, bring the bucket inside after it has snowed and allow the snow to melt. Take your measurements as usual.

Procedure

1. Ask students how much rain and/or snow they think your city gets during the month of ______ (whatever month is coming next). Record their guesses on the board. Be sure to use whatever units or descriptions they give (inches, feet, up to my knees) without correcting or making suggestions.
2. Ask if anyone knows the word meteorologists use for rain, hail, and snow. [precipitation] Have them offer their ideas for how meteorologists could find out the amount of precipitation in a day, week, month, year, etc. Discuss the units meteorologists use to describe the amount of precipitation. [inches and/or feet]
3. Tell students that they are going to use a simple rain gauge to measure the amount of precipitation the school gets for a month (week, all school year).
4. Show students the container you will be using to collect precipitation. Take them outside to the location where the container will be kept. Explain that you are going to pretend that it is a rainy day so that students can get some practice measuring the amount of precipitation.
5. With the watering can, "rain" a few inches of precipitation into the container.
6. Demonstrate how to put the stick into the water so that it touches the bottom of the bucket. Emphasize the importance of holding the stick straight up and down, not at an angle. Remove the stick and have students identify the water mark. If there is no water mark, have students mark the water level with their thumbs.
7. Select a student to use a ruler to measure the amount of rain that collected in the bucket.
8. Dump out the water and repeat this process with different amounts of rain, giving all students a chance to estimate and then to make an accurate measurement.
9. Leave the rain gauge outside and collect data as it becomes available. Ask students to estimate the amount of precipitation before they measure. Record the precipitation amounts on the classroom calendar and/or in students' weather journals.

Connecting Learning

1. What is precipitation? [rain, snow, hail]
2. How much precipitation do you think we get in one week? ...month? ...school year?
3. How can we find out?
4. What tools do we need to measure precipitation?
5. What are you wondering now?

Picturing Precipitation

Topic
Precipitation

Key Question
What are rain, snow, sleet, and hail?

Learning Goals
Students will:
- learn some characteristics of rain, snow, sleet, and hail;
- learn that rain is a liquid, while snow, sleet, and hail are solids;
- draw pictures of each kind of precipitation; and
- show the correct picture when a poem about that kind of precipitation is read.

Guiding Documents
Project 2061 Benchmarks
- *Water can be a liquid or a solid and can go back and forth from one form to the other. If water is turned into ice and then the ice is allowed to melt, the amount of water is the same as it was before freezing.*
- *Draw pictures that correctly portray at least some features of the thing being described.*

NRC Standard
- *Materials can exist in different states—solid, liquid, and gas. Some common materials, such as water, can be changed from one state to another by heating or cooling.*

Science
Earth science
    weather
        precipitation
            rain, snow, sleet, hail
Physical science
    state of matter
        solids, liquids

Integrated Processes
Observing
Comparing and contrasting
Identifying

Materials
Crayons
Card stock (see *Management 1)*
Fabric scraps (see *Management 3)*
Glue
Hole punch (single)
Paper fasteners, one per student
Transparency film, optional (see *Management 2)*

Background Information
Rain, snow, sleet, and hail are four different kinds of precipitation. Each forms in a different way and has different characteristics. When water vapor condenses in the atmosphere, it forms clouds. When the water content of the clouds reaches a certain point, it can no longer stay suspended and falls to the ground as liquid rain. Snow also forms from condensed water droplets, but from high clouds made of ice particles. When these ice crystals fall to Earth through air that is at or below freezing (0°C or 32°F), they reach the ground as snow. Sleet is rain that has frozen between leaving the cloud and reaching the ground. It may also be snow that has partially melted and then refrozen in the air. It is similar in appearance to hail, but is different because of how it forms. Hail forms in clouds when water vapor condenses and then freezes to form balls of ice. These ice balls fall to the ground when they become too heavy to stay in the cloud.

Management
1. Copy the student pages onto card stock and cut them in half along the solid lines. The book cover (window page) should be on colored card stock, and the others should be on white. To make his or her own book, each student needs one set of picture pages and one window page.
2. Before doing the activity, cut out the window on the book cover for each student. If desired, tape a piece of transparency film to the back of the hole to represent the glass or laminate the page after the window has been cut out.
3. Collect and cut to size a variety of fabric scraps that can be glued to the book covers to make curtains. If fabric is not available, paper towels can be substituted.

Procedure

1. Review (or introduce) the word *precipitation*. Ask students to name some of the kinds of precipitation with which they are familiar. Record their responses on the board and add any that they may not have mentioned so that rain, snow, sleet, and hail are all on the board.
2. Ask students how they can tell the difference between these different kinds of precipitation. Write some key words next to each kind of precipitation that help identify it (Rain—wet; Snow—cold, white, soft; Hail—round, hard; Sleet—cold, icy; etc.).
3. Tell students that you are going to read them some riddles and that you would like them to listen carefully and identify each kind of precipitation based on the description given in the riddle.
4. Read the four riddles, one at a time, and have students determine the kind of precipitation described by each.
   - I am wet. I fall in drops from the sky. I am a liquid. What am I?" [rain]
   - I am cold and white. I fall in flakes from the sky. I am a solid. What am I?" [snow]
   - I am cold and white. I fall in hard balls from the sky. I am a solid. What am I?" [hail]
   - I am cold. When rain freezes, I am what forms. I am a solid. What am I?" [sleet]
5. Explain that students are now going to illustrate a book that shows these four kinds of precipitation. Distribute crayons and the four picture pages to students. Read the poems on the pages together as a class.
6. Tell students that on each page they are to draw pictures of themselves and add the precipitation described by the poem. For example, on the rain page, they should add rain in the sky and some puddles on the ground.
7. When students are done illustrating the pages, give each a book cover with the window cut out, fabric scraps, and glue. Demonstrate how to use the fabric scraps to make window curtains, and assist students in gluing their pieces down.
8. After the glue has dried, have students stack their pages so that the window page is on top and the illustrations can be seen through the window. Move from child to child and punch a hole through the top left corner of the pages and fasten them together using a paper fastener.
9. Show students how to spin the illustrated pages out of the way one at a time so that each illustration can be seen through the window. Read the different poems and have students show you their illustrations for each one through the window.

Connecting Learning

1. What are some different kinds of precipitation? [rain, snow, sleet, hail]
2. How did you illustrate rain? Why? ...snow? ...sleet? ...hail?
3. Which was the easiest for you to illustrate? ...the hardest? Why?
4. How can you tell if precipitation is rain, snow, sleet, or hail?
5. How are hail and snow alike? How are they different? How are rain and sleet alike? How are they different?
6. What is the temperature usually like when it snows? ...rains?
7. Does it ever snow in the summer? Why or why not?
8. Is rain a liquid or a solid? [liquid] ...snow? [solid] ...sleet? [solid] ...hail? [solid]
9. What are you wondering now?

Cut out

Picturing Precipitation

Cut out

Picturing Precipitation

Open the curtains and you will see,
it's raining, it's raining,
it's raining on me.
Liquid rain falls in big drops.
Puddles form in front of shops.

Open the curtains and you will see,
it's snowing, it's snowing,
it's snowing on me.
Snow is ice, so soft and white.
Makes the world so cold and bright.

Open the curtains and you will see,
it's hailing, it's hailing,
it's hailing on me.
Balls of ice fall from a cloud.
They hit the roof; it's very loud!

Open the curtains and you will see,
it's sleeting, it's sleeting,
it's sleeting on me.
Frozen rain comes falling down.
Roads are icy in our town.

When the Rain Comes Down

Tune: Do Your Ears Hang Low?

When the snow comes down, it makes piles on the ground.
It makes everything so white; there are snowmen all around.
Snow is solid, don't you know? Yes, you can make a ball to throw.
Oh, when the snow comes down.

When the hail comes down, it's like marbles on the ground.
It makes noises on the roof. There is nothing like that sound.
Hail is solid, don't think twice. It's really little balls of ice.
Oh, when the hail comes down.

When the sleet comes down, it is icy all around.
All the roads get really slick and it bounces on the ground.
Sleet is solid, that is plain. 'Cause it is really frozen rain.
Oh, when the sleet comes down.

Water Cycle Watchers

Topic
Water cycle

Key Question
What are the processes of the water cycle?

Learning Goals
Students will:
- observe a water cycle model,
- identify the processes of the water cycle, and
- relate the model to real-world events.

Guiding Documents
Project 2061 Benchmarks
- *A model of something is different from the real thing but can be used to learn something about the real thing.*
- *When liquid water disappears, it turns into a gas (vapor) in the air and can reappear as a liquid when cooled, or as a solid if cooled below the freezing point of water. Clouds and fog are made of tiny droplets of water.*

NRC Standards
- *Water, which covers the majority of the earth's surface, circulates through the crust, oceans, and atmosphere in what is known as the "water cycle." Water evaporates from the earth's surface, rises and cools as it moves to higher elevations, condenses as rain or snow, and falls to the surface where it collects in lakes, oceans, soil, and in rocks underground.*
- *Materials can exist in different states—solid, liquid, and gas. Some common materials, such as water, can be changed from one state to another by heating or cooling.*

Science
Earth science
weather
water cycle

Integrated Processes
Observing
Comparing and contrasting
Inferring
Relating

Materials
For each water cycle model:
2-L soda bottle, empty and clean
clear plastic cup, 9 oz
hot water in 5-oz Styrofoam cup
ice
tape
scissors
labels and pictures

For each student:
student page
glue stick

Background Information
A soda bottle containing water models the water cycle. Warmth causes the motion of water molecules to quicken; some escape from the *accumulation* in the bottom of the bottle and *evaporate* into the air as invisible water vapor. As the water vapor cools when it hits the cup of ice, the molecules slow down and *condense* into a visible liquid. Water droplets will form and slowly *precipitate* to the bottom of the bottle where they *accumulate.*

Vocabulary labels are included that can be taped to the outside of the soda bottle to help students relate the vocabulary to the processes involved in the bottle. Pictures are also included that will help students relate the processes to the real world. A picture of a lake represents water accumulation. The cloud picture represents condensation. The picture of rain represents precipitation. The process of evaporation cannot be represented because it is invisible. Water vapor cannot be seen.

The cyclic nature of water should be emphasized, noting that the amount of water on Earth does not change. It's location may change, but the amount always remains the same. Some may be falling as precipitation. Some may be stored in oceans or lakes. Some may be in clouds, and some may be in the air.

Management
1. One model can be made for the whole class, or each group can make their own model. If groups are making their own models, cut the necks off the bottles before distributing them.

2. Warm tap water can be used. The warmer it is, the faster the students will see the various components of the water cycle. If student groups are making their own models, make sure the water is not too warm for them to handle.
3. Fill the 9-oz cup two-thirds full of ice.
4. Cut approximately 5 centimeters (2 inches) off the top of the bottle. The resulting hole should be large enough that the 9-oz cup will sit in the top of the bottle, but not large enough that it will fall through. If the hole is too large, roll clay into a rope and place around the cut edge of the bottle. The clay will form a tight seal, preventing the evaporated water from leaving the bottle.

Procedure

1. Hold a class discussion about where the rain comes from, how it gets in the clouds, etc. Lead students to think about the natural water cycle. Talk about the processes of evaporation, condensation, precipitation, and accumulation.
2. Tell students they will observe a model of the water cycle in a soda bottle so they can learn about how it works.
3. Show them the bottle with the top cut off. Tell them that this represents the Earth and the air around the Earth. Pour the warm water in the bottom of the bottle. Tell students that this represents water that has accumulated in a lake or an ocean. Let them know that the sun has warmed the water.
4. Next, put the cup of ice in the opening of the soda bottle. Tell students that the reason you put ice in the cup is because air high in the sky is cooler.
5. Invite the students to wait awhile and then describe what they see. (They should see moisture condense on the cup that contains ice.) Tell the students that the moisture they see represents clouds in the sky. Inform them that the water *condenses* because it has cooled.
6. Wait until the water starts to drip off the cup. Tell the students that this is *precipitation* that falls from the clouds. Often it is rain, but sometimes it is snow or hail or sleet. Make sure that they understand that it accumulates again and will continue the cycle.
7. Cut out the labels of the processes and tape them to the bottle. Also cut out the related pictures and tape them on the bottle.
8. Discuss how we cannot see *evaporation*, but it happens when water turns to a gas in the air.
9. Distribute the student page. Direct students to cut out the labels and pictures and glue them onto the picture of the water cycle bottle.

Connecting Learning

1. Where did the water accumulate in our water cycle bottle? [in the bottom]
2. Where does water accumulate in the real world? [in oceans, lakes, ponds, etc.]
3. Where does water condense in our water cycle bottle? [around the cold cup]
4. Where does it condense in the real world? [in the clouds]
5. How did our water cycle model show us precipitation? [Drips fell off the cold cup.]
6. What kind of precipitation falls on Earth? [rain, snow, sleet, hail]
7. Why didn't we see the water evaporate? [It is invisible.]
8. Where is evaporated water found? [in the air]
9. Why do you think it is important that the Earth has a water cycle? [All living things need water to live. Without the water cycle, we would run out of water.]
10. What are you wondering now?

Water Cycle Watchers

Labels and Pictures for the Model

Evaporation

Precipitation

Condensation

Accumulation

Water Cycle Watchers

Cut off this half of the page. Put the pictures and the labels on the picture of the water cycle bottle.

Evaporation
Condensation
Precipitation
Accumulation

Round and Round It Goes

Topic
Water cycle

Key Question
What is the water cycle?

Learning Goals
Students will:
- explain how the water on Earth moves in a continuous cycle,
- identify the stages in the water cycle, and
- model the water cycle.

Guiding Documents
Project 2061 Benchmark
- *When liquid water disappears, it turns into a gas (vapor) in the air and can reappear as a liquid when cooled, or as a solid if cooled below the freezing point of water. Clouds and fog are made of tiny droplets of water.*

NRC Standard
- *Materials can exist in different states—solid, liquid, and gas. Some common materials, such as water, can be changed from one state to another by heating or cooling.*

Science
Earth science
weather
water cycle

Integrated Processes
Observing
Recording
Relating
Generalizing

Materials
For the class:
precipitation (see *Management 1)*
yarn
water cycle signs (see *Management 2)*
arrows (see *Management 3)*

For each student:
water cycle book (see *Management 4)*

Background Information

The water cycle can briefly be described as the continuous movement of the Earth's water from its surface, to the air, and back to its surface. The heat from the sun evaporates the water on the Earth's surface. Cool air condenses the water vapor into water droplets or ice crystals. They fall back to the Earth as precipitation. Around 75% of the precipitation falls directly into the oceans. The rest evaporates immediately or may soak into the Earth and become part of the ground water supply. Much of the surface and ground water eventually returns to the oceans (accumulation), beginning the entire process over again.

Management
1. Miniature marshmallows or Styrofoam peanuts will work well as precipitation.
2. Make copies of the condensation, evaporation, accumulation, and precipitation pages on card stock. Laminate for extended use. Tape the ends of a length of yarn to the top corners of the condensation, evaporation and precipitation pages so that students can wear them like necklaces. The accumulation page will be placed on the ground where the precipitation is collecting.
3. Copy three arrows for *Part Two* closure.
4. To make a water cycle book, copy the five pages and staple along the left edge.
5. The student book is written at a second grade reading level; however, you may choose to read it aloud together.

Procedure
Part One
1. Display a glass of water. Have students look at the water, then ask if anyone can guess how old it is. Explain that the water in the glass may have fallen from the sky as rain just last month, but the water itself has been around pretty much as long as the Earth has. That means that when the dinosaurs roamed, the water in the glass could have been part of lakes from which they drank.
2. Tell students that the Earth has a limited amount of water, and that water keeps going around and around and around in what we call the water cycle. Explain that the cycle is made up of four parts: evaporation, condensation, precipitation, and accumulation.

3. Tell students that they are going to be reading about each of the parts of the water cycle. Distribute the water cycle book. Allow time for students to read. Discuss what each part of the water cycle is called, the order of the cycle, and anything that they learned about the water cycle through the story.
4. Give students the page with the model of the water cycle and have them label each part with the correct letter.

Part Two

1. Ask the students what they remember about water and the cycle that it goes through.
2. Tell the class that they are going to get a chance to act out the water cycle.
3. Assign one student to wear the condensation necklace, one student to wear the evaporation necklace, and one to wear the precipitation necklace.
4. Place several marshmallows/Styrofoam pieces on the ground. Bring students' attention to the "accumulation" of precipitation on the ground. Explain that each represents a drop of water.
5. Ask the class what happens to the accumulated water in the water cycle. [The sun warms the water and the drops of water evaporate into the air.] Ask the student playing the sun (evaporation) to begin evaporating or picking up the drops of water and handing them to the cloud.
6. Have the student predict what will happen when the cloud can no longer hold the drops of water. [They will fall back to Earth.] As they begin to fall, ask the student wearing the precipitation necklace to move in and tell about all of the different forms the water could take as it falls.
7. Repeat procedures four through six several times, allowing different students to play the various parts of the water cycle.
8. Close the lesson by asking the students to help you place the water cycle signs in the correct order on the board. Invite a student to place arrows between the signs to show where the water is going next.

Connecting Learning

1. Could we be drinking water that a dinosaur swam in? Explain.
2. Does the Earth have more or less water now than 1000 years ago? [The amount of water on Earth is the same.]
3. What is evaporation? [when water changes from a liquid to a gas] What in our water cycle was responsible for evaporating the water from the Earth? [the sun]
4. Can you normally see evaporation? Explain. [No, water vapor is an invisible gas.]
5. What is condensation? [when water vapor cools and changes back to a liquid] ...precipitation? [rain, snow, sleet, hail, etc.] ...accumulation? [precipitation that has collected/gathered in a river, lake, ground water, etc.]
6. How do all the parts of the water cycle fit together?
7. What would happen if one part were left out?
8. What are you wondering now?

1

Evaporation

The sun heats the water in lakes, rivers, and the ocean. It turns the water into vapor. The water vapor goes into the air.

2

Condensation

Water vapor in the air gets cold. It changes back into liquid. The liquid makes clouds. This is called condensation.

3

Precipitation

Precipitation happens when a lot of water has condensed. When the clouds cannot hold it all, water falls back to the Earth. The water comes down as rain, hail, sleet, or snow.

4

Accumulation

When water falls back to Earth, it might fall back in the oceans, lakes, or rivers. It may end up on land. When it falls on land, it may soak into the Earth. It becomes part of the water that plants and animals use to drink. It may run over the soil and accumulate in the oceans, lakes, or rivers. Then the cycle repeats by evaporating the water.

5

Water Cycle

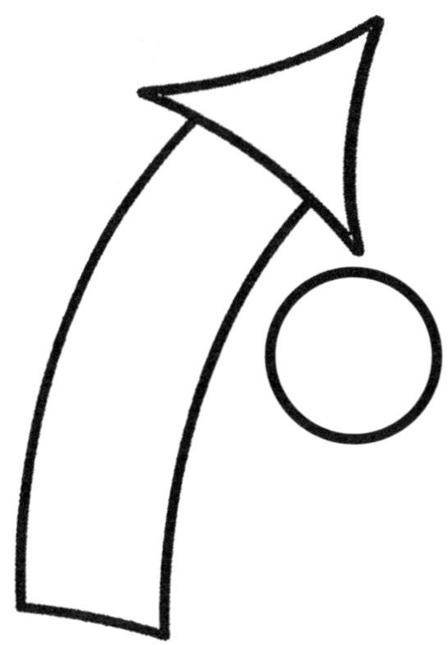

A= Evaporation
B= Condensation
C= Precipitation
D= Accumulation

Label the circles with each part of the water cycle.

Condensation

Evaporation

Precipitation

Accumulation

Reader's Theater: The Water Cycle

Topic
Water cycle

Key Question
What can we learn about the water cycle from a reader's theater?

Learning Goal
Students will learn about the water cycle by listening to and/or participating in a reader's theater.

Guiding Documents
Project 2061 Benchmarks

- *When liquid water disappears, it turns into a gas (vapor) in the air and can reappear as a liquid when cooled, or as a solid if cooled below the freezing point of water. Clouds and fog are made of tiny droplets of water.*
- *The sun warms the land, air, and water.*
- *Water can be a liquid or a solid and can go back and forth from one form to the other. If water is turned into ice and then the ice is allowed to melt, the amount of water is the same as it was before freezing.*

NRC Standard

- *Materials can exist in different states—solid, liquid, and gas. Some common materials, such as water, can be changed from one state to another by heating or cooling.*

Science
Earth science
weather
water cycle

Integrated Processes
Observing
Relating
Interpreting

Materials
Character hats and necklaces (see *Management 1)*
Yarn
Tape
Reader's theater scripts

Background Information

Water on Earth is constantly moving through the water cycle. Liquid water accumulates in rivers, lakes, and oceans. From there, heat from the sun causes it to evaporate into the air. When the water vapor cools to a certain point, it condenses, forming clouds. Eventually, the water returns to Earth in the form of precipitation.

Using the format of a reader's theater allows students to explore the water cycle in a way they may never have experienced. By reading and/or listening to the story, they become engaged in the action as they follow two water drops on a journey through the water cycle. The script reinforces the ongoing nature of the water cycle and the fact that a single drop of water will take many different paths throughout the course of time.

Management

1. Copy all the student pages onto card stock except for the scripts. Laminate the pages if possible. For the sun, water drops, and narrator, cut out the art and tape on the included strips to make each into a hat. These can then be adjusted to fit students' heads. The snowflakes will be taped onto the raindrops and then removed during the course of the reader's theater.
2. Make two card stock copies each of the *Cloud* and *Water Vapor* pages and turn them into necklaces by taping a length of yarn to the upper corners of each page.
3. You may wish to have students sit on stools or chairs while performing the reader's theater. You may also wish to provide them with music stands on which to place their scripts so that their hands are free to gesture and change "costumes" as needed.
4. Select four students to perform the reader's theater. If you have time, multiple groups can be given the opportunity to perform. Students should be given time to read the script and rehearse together before presenting to the class. This may need to happen over a period of several days.

Procedure

1. Distribute the scripts and character hats and necklaces to the selected students. Allow them time to read the scripts and rehearse together.

2. When students are ready to perform, have the readers come to the front of the class while you set the scene. Tell students that they are going to be watching a reader's theater. It's kind of like a play, but the readers don't memorize the lines or act things out in the same way.
3. Explain that this reader's theater is about the water cycle. Introduce each of the four characters: the narrator, the sun, and the two water drops. Invite students to listen quietly as their classmates perform.
4. If desired, repeat with other groups of students who have also prepared.
5. Discuss what students learned about the water cycle and how the story illustrated the different phases of the water cycle.

Connecting Learning

1. How did the two water drops change during the story? [They changed from liquid, to water vapor, to a cloud, to snow, and back to liquid again.]
2. What caused their changes? [The sun caused them to evaporate. Cool air caused them to condense and fall as snow. The sun caused them to melt and become liquid again.]
3. In what phase of the water cycle did the water drops begin the story? [accumulation] How do you know? [They were in a lake.]
4. What phase of the water cycle were the water drops in when they were part of a cloud? [condensation] ...when they were snowflakes? [precipitation] ...when they were water vapor? [evaporation]
5. What did you like most about this reader's theater?
6. What are you wondering now?

Extensions

1. Have students write and perform their own reader's theaters about the water cycle or other related topics.
2. Read *A Drop Around the World* and compare the journey of the raindrop in that story to the one of those in the reader's theater.

Curriculum Correlation

McKinney, Barbara Shaw. *A Drop Around the World.* Dawn Publications. Nevada City, CA. 1998.

Reader's Theater: The Water Cycle

Characters: Narrator, Water Drop 1, Water Drop 2, Sun

Narrator: Our story begins as the sun is watching two drops of water playing in a lake.

Water Drop 1: It's getting hot. I don't think I can swim any more. I am getting light headed. I feel as light as air. The sun is doing it again!

Sun: I can't help it. I'm hot and full of energy. I heat the land and water.

Water Drop 2: I feel funny! I'm getting dizzy too! I cannot swim any longer. Maybe I'll just float for a while. *(Put on Water Vapor necklace.)*

Water Drop 1: Oh no! You're not floating in water. You are floating in the air! You're not a drop of water. You're water vapor. I can't see you.

Water Drop 2: What's water vapor?

Water Drop 1: It is water that is heated. It turns into a gas. We are both floating up. We are evaporating. *(Put on Water Vapor necklace.)*

Water Drop 2: There are more drops of water turning into water vapor. Let's join the crowd. *(Put on Cloud necklace.)*

Water Drop 1: Don't you mean the cloud? Okay, let's condense. *(Put on Cloud necklace.)*

Water Drop 2: What does that mean?

Water Drop 1: Condense means that we'll change back into liquid water. Then we'll be part of a cloud.

Narrator: Our water drops join many others. Together they make a fluffy cloud. The air gets colder and colder. The cloud is getting heavy. The air cannot hold it all. Precipitation falls back down to Earth. It's about to snow!

Water Drop 2: *(Remove necklaces. Tape snowflake over water drop.)* Hey, what are cold, white, and one of a kind?

Water Drop 1: *(Remove necklaces. Tape snowflake over water drop.)* Don't be a flake. Wait, I guess we're both flakes—snowflakes, that is.

Water Drop 2: Help we're falling! Hey, where are you going?

Water Drop 1: Down.

Narrator: The two snowflakes fell gracefully down. They landed on a mountain.

Water Drop 1: Well, it looks like we're stuck here. Now we wait.

Water Drop 2: What are we waiting for?

Water Drop 1: We have to wait until we melt. This is NOT my favorite part of the water cycle.

Narrator: A long, long time later, two drops of water roll down the mountain and into a river.

Water Drop 1: *(Remove snowflake.)* Wow, we've finally melted.

Water Drop 2: *(Remove snowflake.)* We are free at last!

Water Drop 1: Watch out for that rock!

Water Drop 2: Where are we? This isn't our lake.

Water Drop 1: We are on our way to the ocean.

Water Drop 2: The ocean? I thought we would just fall back into our lake.

Water Drop 1: I've taken this trip a million times, and every time it's different.

Water Drop 2: Where has this cycle taken you?

Water Drop 1: I have been well water in Wisconsin, a typhoon in Thailand, and rain in Spain. Once I was even a snowflake that fell on the nose of a tyrannosaurus rex.

Sun: Hi. Where have you two been? I missed you.

Water Drop 1: Don't ask. We are just glad to be floating in this peaceful, cool gulf water. There sure are a lot of water drops accumulating around here.

Water Drop 2: Is it getting hot in here?

Water Drop 1: Oh no, not again!

Water Drop 2: I wonder where we'll end up this time?

Copy this page onto card stock and cut out the sun. Tape one hat strip to each side of the sun so that the slits are facing in opposite directions.

Copy this page onto card stock and cut out the water drop. Tape one hat strip to each side of the water drop so that the slits are facing in opposite directions.

Copy this page onto card stock and cut out the water drop. Tape one hat strip to each side of the water drop so that the slits are facing in opposite directions.

Copy this page onto card stock and cut out the top hat. Tape one hat strip to each side of the top hat so that the slits are facing in opposite directions.

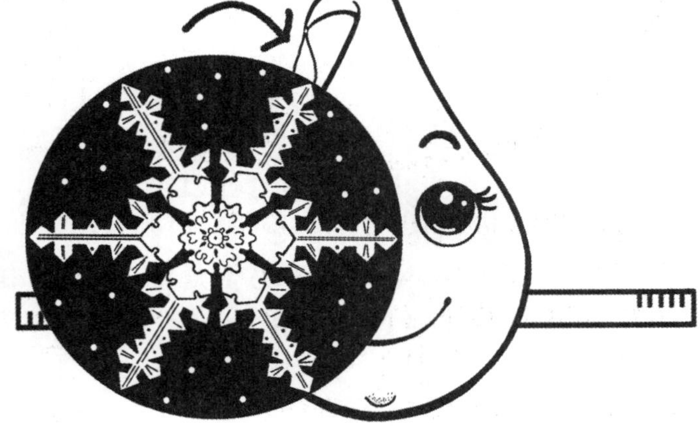

Copy this page onto card stock and cut out the snowflake. Attach a loop of tape to the back so that it can be affixed to the water drop.

Copy this page onto card stock and cut out the snowflake. Attach a loop of tape to the back so that it can be affixed to the water drop.

Cloud

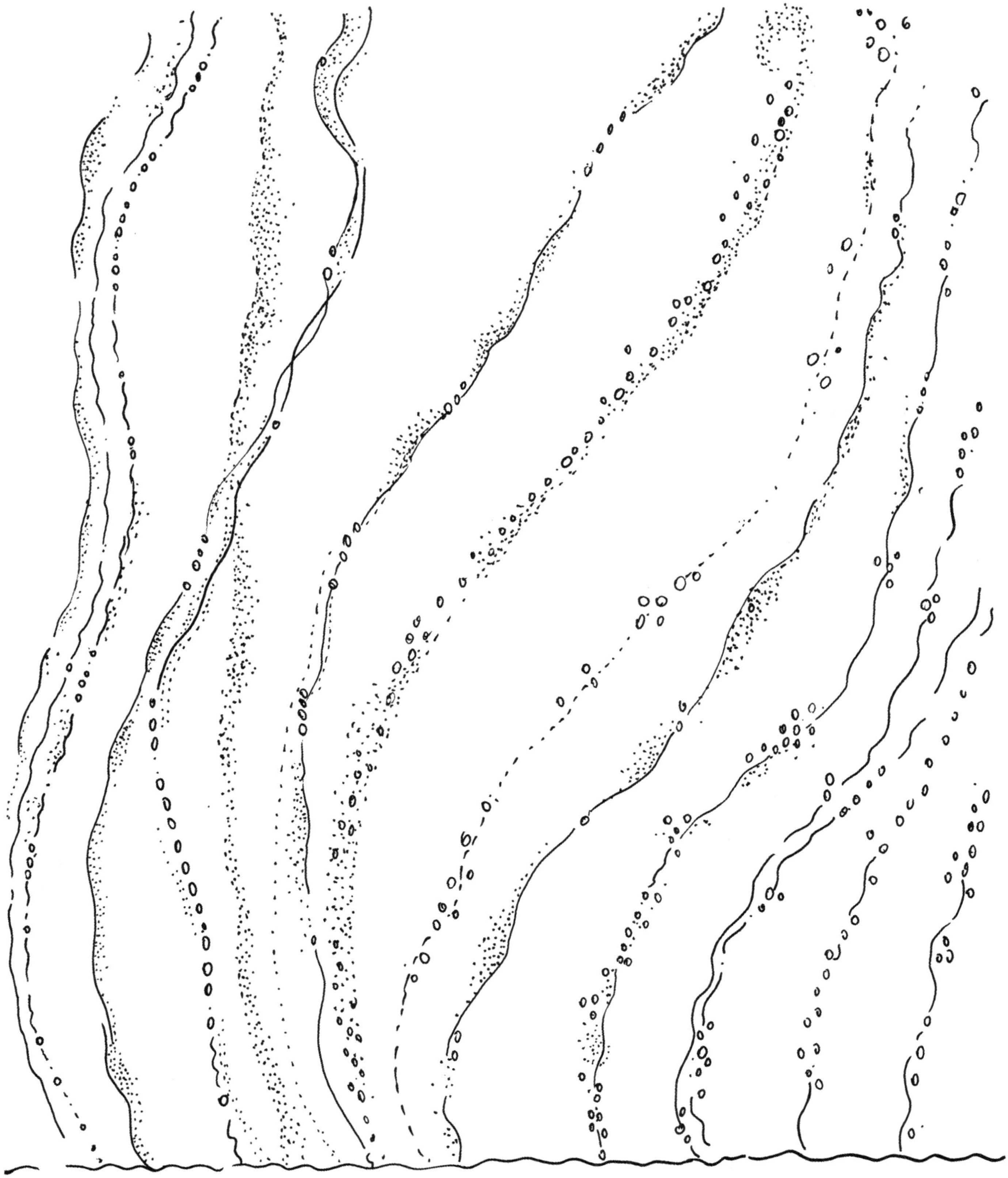

Water Vapor

Reader's Theater: The Water Cycle

Hat Strips

Cut these hat strips apart and cut along each of the slits.

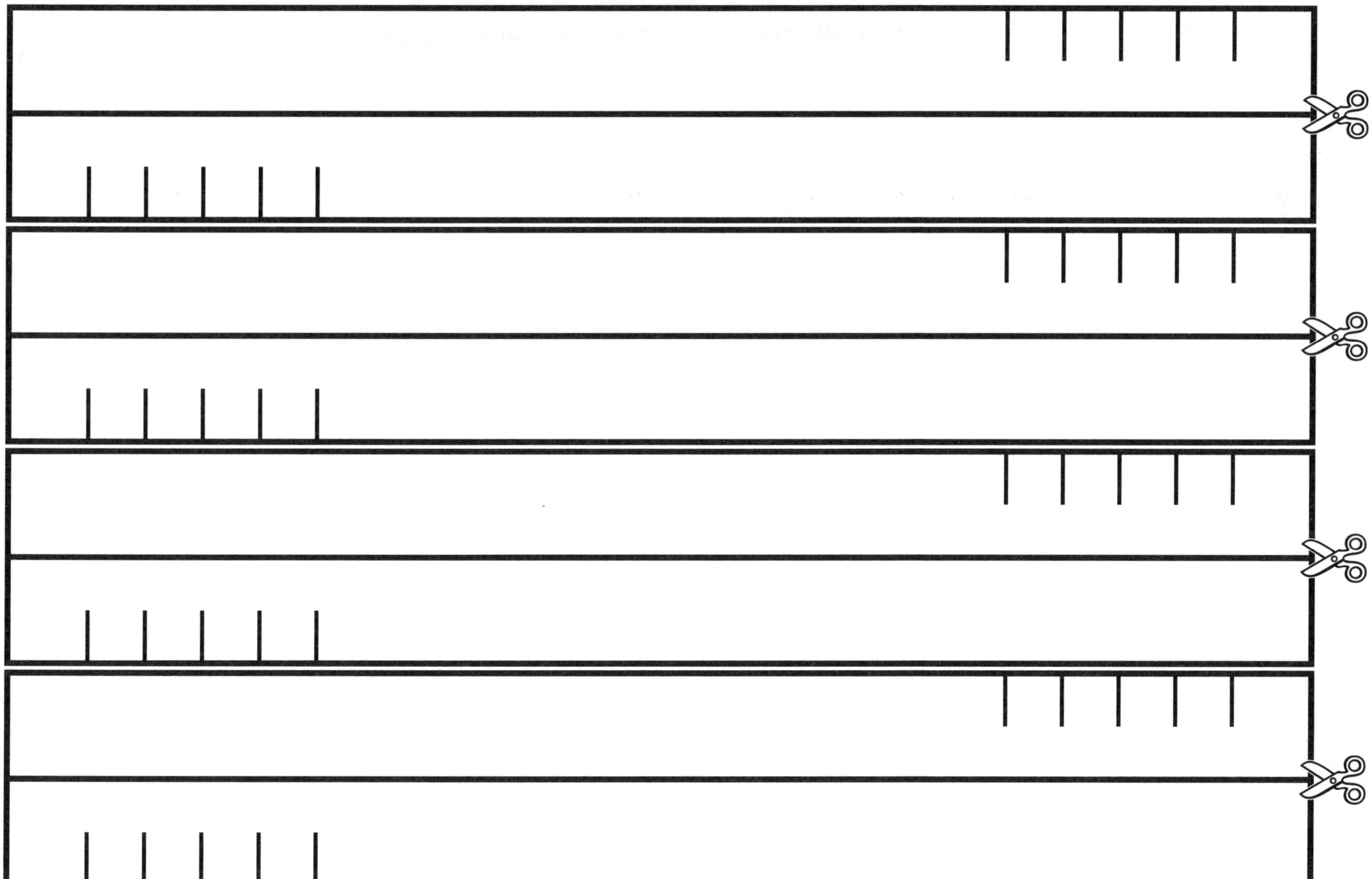

Water Cycle Song

Words by Suzy Gazlay

Tune: Clementine

PUDDLE PICTURES

Topic
Evaporation

Key Question
What happens to the pictures we paint on the sidewalk?

Learning Goals
Students will:
- make careful observations, and
- discover that something happens to the water when they try to paint a picture outside on the sidewalk.

Guiding Documents
Project 2061 Benchmarks
- *People use their senses to find out about their surroundings and themselves. Different senses give different information. Sometimes a person can get different information about the same thing by moving closer to it or further away from it.*
- *Raise questions about the world around them and be willing to seek answers to some of them by making careful observations and trying things out.*
- *When liquid water disappears, it turns into a gas (vapor) in the air and can reappear as a liquid when cooled, or as a solid if cooled below the freezing point of water. Clouds and fog are made of tiny droplets of water.*

NRC Standards
- *Employ simple equipment and tools to gather data and extend the senses.*
- *Communicate investigations and explanations.*
- *Materials can exist in different states—solid, liquid, and gas. Some common materials, such as water, can be changed from one state to another by heating or cooling.*
- *Water, which covers the majority of the earth's surface, circulates through the crust, oceans, and atmosphere in what is known as the "water cycle." Water evaporates from the earth's surface, rises and cools as it moves to higher elevations, condenses as rain or snow, and falls to the surface where it collects in lakes, oceans, soil, and in rocks underground.*

Science
Physical science
  states of matter
Earth science
  weather
    water cycle
      evaporation

Integrated Processes
Observing
Predicting
Comparing and contrasting
Collecting data
Communicating
Drawing conclusions

Materials
A container of water for each student or small group
Paintbrush for each student (see *Management 3)*

Background Information

In this activity, students are invited to experiment with the concept that water exists as a liquid and as a gas. They may not be able to understand evaporation, but simple explorations allow them to observe the phenomenon—the water "disappears." It is important that teachers listen carefully to students' observations and plan for other investigations to further explore the concept and dispel misconceptions that may exist.

Management
1. Choose a blacktop or sidewalk area with full exposure to the sun to be used for painting water pictures.
2. Gather wide-bottomed containers for the water so they will not easily tip over.
3. Each child will need a long-handled paintbrush. One-inch to two-inch wide student brushes work well.
4. This activity is best done on a warm, sunny day with low humidity. After the initial investigation, you may want to try painting on days when it is cloudy, humid, windy, foggy, hot, etc.

Procedure

1. Present students with the challenge to use water to paint a picture on the sidewalk before the picture disappears.
2. Take students to the area of the playground to be used for the activity. Discuss how the experiment will be conducted and listen to students' ideas about the painting that they will do.
3. Describe and model painting a picture with water.
4. Explain that each student will have the opportunity to paint his or her own puddle picture. Remind students that each of them should paint his or her own picture, taking care not to interfere with any one else's picture.
5. Urge students to pay special attention to what happens to the water as they paint. Distribute cups of water and paintbrushes.
6. After all students have had time to paint a picture with water and observe what happens, gather them together to share their observations and discoveries.

Connecting Learning

1. How did you paint your puddle picture?
2. What happened to your picture as you painted? [It disappeared.]
3. Did all of us have the same thing happen to our picture? Why do you think this happened?
4. What part of the water cycle does this represent—evaporation, condensation, precipitation, or accumulation? [evaporation]
5. Can you think of any other time you have noticed water doing what it did as you painted? Explain. [Rain puddles dry up. The clothes that are hung out on the clothesline dry. When we go swimming and dry off with a towel, the towel dries.]
6. Was there anything that you noticed that surprised you? What did you think would happen? What happened?
7. Is there anything we could do to help us save our puddle pictures?
8. What other ways could we experiment with water?
9. What are you wondering now?

A Disappearing Act

Topic
Evaporation

Key Question
What happens to moisture on the playground and on the table top?

Learning Goal
Students will see the results of water disappearing (evaporating) into the atmosphere.

Guiding Documents
Project 2061 Benchmarks

- *People can often learn about things around them by just observing those things carefully, but sometimes they can learn more by doing something to the things and noting what happens.*
- *Water left in an open container disappears, but water in a closed container does not disappear.*
- *The sun warms the land, air, and water.*
- *Tools such as thermometers, magnifiers, rulers, or balances often give more information about things than can be obtained by just observing things without their help.*

NRC Standards

- *Ask a question about objects, organisms, and events in the environment.*
- *Employ simple equipment and tools to gather data and extend the senses.*

*Common Core Standards for Math**

- *Measure and estimate lengths in standard units. (2.MD)*
- *Work with time and money. (2.MD)*

Math
Measurement
  time
Estimation

Science
Earth science
  weather
    water cycle
      evaporation

Integrated Processes
Observing
Comparing and contrasting
Collecting and recording data
Predicting
Inferring
Generalizing

Materials
Balance
Two identical jars with lids
Water
Chalk
Cloth or paper towel
Ruler (see *Management 4)*
Stopwatch, optional

Background Information

Evaporation is the process by which a liquid changes to a gas. When water evaporates, it changes to an invisible, odorless gas called water vapor. Changes in states of matter require a transfer of energy. Energy from the sun causes this evaporation.

Students are not able to witness the evaporative process; they can, however, observe the results of evaporation. In this activity, students will observe puddles on the playground and notice that the water disappears; it goes somewhere. They will also observe the disappearance of moisture on a table top. They will collect data by determining the time it takes for the moisture to evaporate.

Management

1. This activity has three parts. *Part One* has students observe puddles on the playground. *Part Two* takes place indoors with students observing the disappearance of the moisture from a table top and from their hands. *Part Three* lets students observe what happens to a balance as water evaporates.
2. The puddles that are being observed on the playground need to be in an area that will not let the water soak into the surface. A concrete sidewalk or paved play area works well.
3. If the weather has not cooperated with rain, make some puddles by pouring water into a depression in the surface being observed.
4. If students do not know how to use rulers, substitute non-customary units such as Unifix cubes. String can also be cut to represent the measure of the puddle.

Procedure

Part One

1. Take the students out on the playground after a rain to observe the puddles. Have the students use chalk to trace around the puddles. Watch the puddles for a period of time. Have the students draw a "before" view of the puddle.
2. Have students measure the diameter of the puddle. Many puddles will not be symmetrical, so have students determine how they are going to measure them and include their methods in their illustrations.
3. Periodically check the puddles to see how much evaporation has occurred. When you feel that it is substantial enough to make an impact on the students, invite them back outside to illustrate and measure their puddles once again. Have the students measure and draw an "after" view of their puddles.
4. Have the students explain what they think happened to the water in the puddles.

Part Two

1. Tell students that they are going to once again watch water disappear. Wipe a damp cloth across a table top. Ask them to explain what they think is happening.
2. Again wipe a small area and draw a circle around it. Watch the moisture disappear. Have the students count, or time with a stopwatch, how long it takes the water to disappear. Discuss with the students what is happening and where the moisture may be going.
3. Tell students to wash their hands but not dry them. Have them time how long it takes for their hands to dry. Ask them to explain what things might cause their hands to dry more rapidly ...more slowly.
4. Discuss how the disappearance of water from the table and from their hands is like the water in the puddle.
5. Ask students to brainstorm other things they could do to observe that water disappears.

Part Three

1. Fill two identical containers with the same amount of water. Put lids on both jars. Place them in the balance and make certain that it is equalized. Remove the lid from one jar and put in the pan so the mass will remain the same. Leave the water and jars intact and observe over a period of several days.
2. Watch what happens to the balance and the water level in the jars. Discuss with the students why the balance is tipping. Have the students draw and write about what they see happening on the *Covered and Uncovered* recording sheet.
3. Ask students to apply the disappearance of water from the jars to the experience with the puddles.

Connecting Learning

1. When a puddle dries up, where does the water go? How do you know?
2. What happened to the wet place on the table? Where did the water go?
3. How can there be water in the air? [The drops are so tiny we cannot see them.]
4. Is there water in the air that we can see? [clouds, fog]
5. Could you see the water evaporate? Why not?
6. What would happen if there were twice as much moisture in the spot on the table or floor? Would it take twice as long to evaporate?
7. What are you wondering now?

Extensions

1. *Parts One* and *Two* of this activity could be repeated on a very humid day and on a dry, sunny day. Results should be compared and discussed.
2. Fill several different-sized containers about one-half full of water. Try to pick containers that have different-sized openings. With a black marking pen, mark the level of the water at the beginning. Watch the level of the water for several days, mark the level of the water daily. Does the size of the opening make any difference in the evaporation of the water?

Curriculum Correlation

Cole, Joanna. *The Magic School Bus at the Waterworks.* Scholastic, Inc. New York. 1986.

Peters, Lisa Westberg. *Water's Way.* Arcade Publishing. New York. 1991.

A Disappearing ACT

1. Go outside. Observe a puddle of water. Go back later.

Draw the puddle.

Now	Later
Measure: ____________	Measure: ____________

What happened to the water in the puddle?

2. Wipe an area with something wet. How long will it take for the water to disappear?

Try 1: I think ________ . I counted ________.

Try 2: I think ________. I counted ________.

Try 3: I think ________ . I counted ________.

3. Where do you think the water goes?

A Disappearing ACT

1. Make two jars of water just alike.

2. Put them in a balance. Make sure both sides are equal.

3. Take the lid off one jar. Put lid in the balance pan.

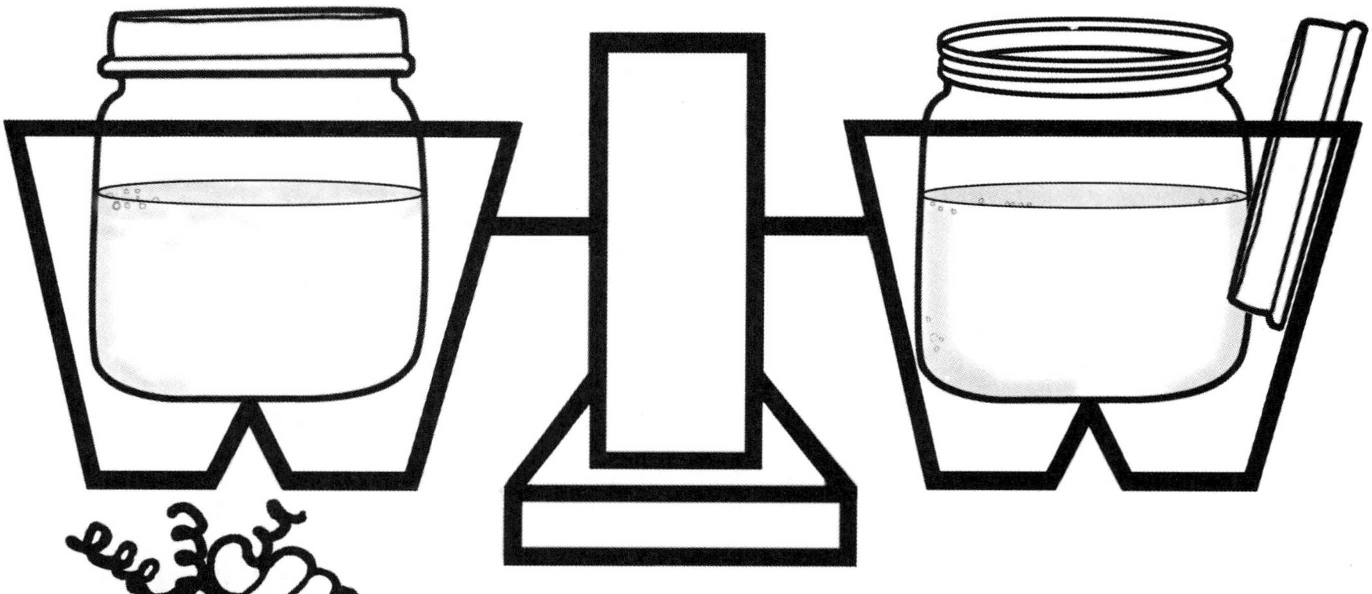

4. Leave it alone for several days. What do you think will happen?

5. Draw and write about what happened.

PACKED and PREPARED

Topic
Disaster preparedness

Key Question
What can you do to prepare for severe weather events and natural disasters, and why is it important?

Learning Goals
Students will:
- discuss the reasons for preparing for severe weather events and natural disasters, and
- identify items that belong in a disaster supply kit.

Guiding Document
NRC Standard
- *Safety and security are basic needs of humans. Safety involves freedom from danger, risk, or injury. Security involves feelings of confidence and lack of anxiety and fear. Student understandings include following safety rules for home and school, preventing abuse and neglect, avoiding injury, knowing whom to ask for help, and when and how to say no.*

Science
Earth science
    severe weather
Safety
    disaster preparedness

Integrated Processes
Observing
Comparing and contrasting
Communicating
Relating

Materials
Construction paper, 9" x 12"
Markers
Scissors
Glue sticks
Suitcase template (see *Management 1)*
Student pages

Background Information
Weather-related natural disasters such as floods and hurricanes can cause tremendous damage and devastation. Preparing for the possibility of severe weather events is important, especially for those living in areas prone to certain kinds of disasters like hurricanes and tornadoes. The Federal Emergency Management Association (FEMA) recommends that families develop a "just in case" plan and an emergency supply kit in case of disaster. This kit should contain the supplies needed to survive for several days, including non-perishable food and clean drinking water, as well as things like a weather radio, first aid kit, clothing, and flashlight.

This activity gives students the opportunity to recognize the need for preparing for emergencies and to discuss the things that are most and least important for survival in case of a natural disaster.

Management
1. Make several copies of the suitcase template on card stock. Cut out the shape along the outer lines to make templates that students can use to trace their suitcases.
2. Do not use black construction paper—students need to be able to write on it with markers.

Procedure
1. Ask students to think of some severe weather events you have experienced (or might experience) in your area. Discuss what effects these events have on people.
2. Talk with students about the importance of being prepared for these kinds of events. Invite their suggestions for ways to be prepared.
3. If no one suggests it, mention the idea of a disaster supply kit. Have students share their thoughts about the kinds of things that might go in such a kit.
4. Explain that students will be making a disaster supply kit by filling a suitcase with the things they think are most important. They will have more things to take than can fit in the suitcase, so they will have to decide which items are the most important.

5. Distribute the construction paper, scissors, and suitcase templates. Instruct students to fold their papers in half widthwise and to take turns tracing the corners and handle of the suitcase on their papers. The bottom of the suitcase should align with the folded edge of the paper and it should fill the width.
6. Give students glue sticks and the page representing the inside of the suitcase. Have them fold it in half along the dashed line so that the printed side is on the inside. Instruct them to glue the page to the inside of the suitcase so that it's centered. They can trim any edges of the paper that extend beyond the edge of the suitcase. Have students use the markers to write "Disaster Supply Kit" on the outside of the suitcase.

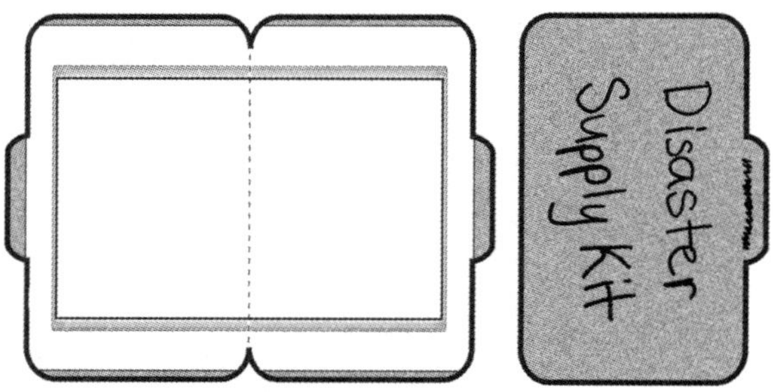

7. Distribute the page of picture cards to each student and have them cut the cards apart. Explain that they will only be able to take what fits in the suitcase. The picture cards cannot overlap. Have students experiment with putting cards into the suitcase until everyone agrees that only eight of the 12 items can be packed.
8. Have students work individually to select the items for their suitcases and glue them down. Students should not share their thoughts or ideas with others at this point; each student should decide what to pack. Remind students that there is not one correct way to pack the suitcase, so they shouldn't worry about getting a "right" answer; they should pack what they think is best.
9. When everyone has had time to pack their suitcases, invite the class to share the items they selected. Discuss each item and have students share whether they think it is essential, important, or just nice to have. Have them justify their responses and explain why they included the items they did. (See *Solutions* for a discussion of the items.)
10. Talk again about the importance of being prepared for severe weather and other natural disasters and any other things students and their families can do to be ready in case of an emergency.

Connecting Learning

1. What are some severe weather events we have experienced? How have those events affected our city?
2. What are some things you can do to be prepared in case of an emergency? [Have a family disaster plan. Have a disaster supply kit.]
3. What things did you choose to put in your kit? What were your reasons for your selections?
4. How did your kit compare to those of others in the class?
5. Were there some items that everyone packed? [Hopefully everyone will have packed the essentials—food and water.] ...no one packed?
6. Is it okay to have items in the kit that are not essential or important? [Yes, as long as you also have the things you need. It can be good to have things to do when you are away from home or without power.]
7. What other supplies do you think it might be important to have in a disaster supply kit? Explain.
8. What else can you do to be ready for emergencies?
9. What are you wondering now?

Solutions

The items shown on the cards are discussed here. There are no right and wrong answers when it comes to how to fill the suitcase. Some things are more important than others, and students should recognize those things that are essential, important, and just nice to have. (This information is from the FEMA for Kids website.)

Water: Essential—One gallon per person per day is recommended.

Food: Essential—Pack non-perishable, ready-to-eat items such as canned soups, meats, and vegetables as well as high-energy foods like peanut butter and granola. Don't forget a can opener!

Clothing: Important—Each person needs a complete change of clothes and good, sturdy shoes. Depending on climate and location, additional clothes may be needed.

First Aid Kit: Important—Basic medical supplies to help deal with small injuries are good to have. Include pain relievers, medication, and vitamins.

Radio: Important—A battery-operated radio (with extra batteries) will help you know what is going on in case of a power outage. There are special NOAA weather radios designed specifically for alerting people to severe weather events.

Flashlight: Important—Power is often lost during severe weather.

Whistle: Important—A whistle can be used to signal rescuers for help.

Tools: Important—In case of power loss, it may be necessary to turn off utilities.

Bleach: Important—Bleach can be used as a disinfectant. It can also be used to purify contaminated water (16 drops per gallon). Don't use color-safe bleach or bleaches with added scents or cleaners.

MP3 Player, Personal Video Game Player, Toys: Nice to have—These items are not essential, but it helps to pass the time if you have things to do.

Internet Connections

FEMA for Kids

http://www.fema.gov/kids/

This government website has lots of information geared toward kids about how to prepare for disasters. Interactive games, coloring pages, and lesson plan ideas can also be found there.

Home Link

Encourage students to take their disaster supply kits home and talk to their parents about the need for such a kit.

Suitcase Template

Trace around the edge of the suitcase.
Align the bottom edge with the fold on your paper.

align with fold

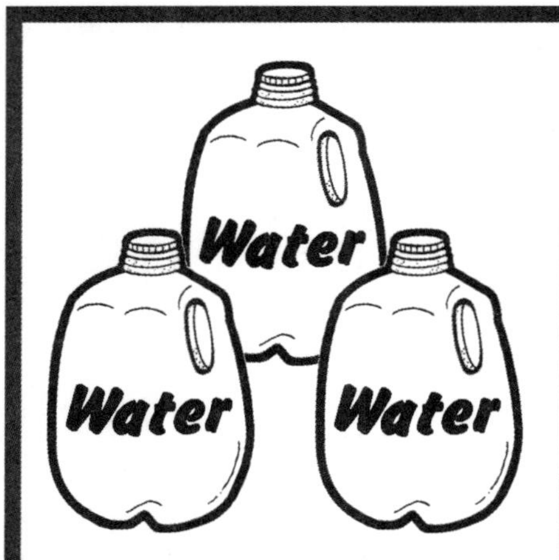
Water
Water
Water

TUNA
TUNA
Ham
Baked
BEANS
Peanut
Butter

FIRST AID

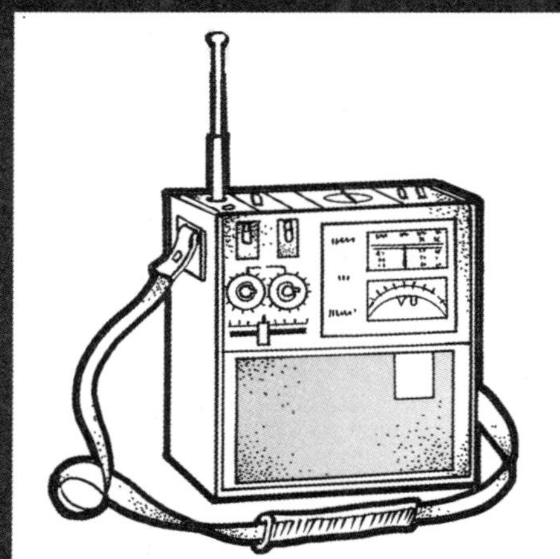

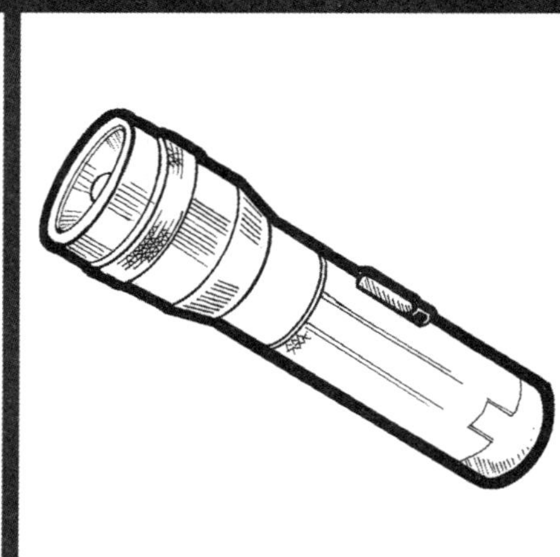

Household
BLEACH

GAME
SHOW

Severe Weather Safety

Topic
Severe weather safety

Key Question
What are the precautions I need to take in order to be safe during severe weather?

Learning Goals
Students will:
- identify severe weather events,
- brainstorm safety precautions that need to be taken during these events, and
- play a game in which they match pictures of safety precautions to the descriptions of the precautions.

Guiding Documents
Project 2061 Benchmark
- *Some events in nature have a repeating pattern. The weather changes some from day to day, but things such as temperature and rain (or snow) tend to be high, low, or medium in the same months every year.*

NRC Standards
- *Safety and security are basic needs of humans. Safety involves freedom from danger, risk, or injury. Security involves feelings of confidence and lack of anxiety and fear. Student understandings include following safety rules for home and school, preventing abuse and neglect, avoiding injury, knowing whom to ask for help, and when and how to say no.*
- *Weather changes from day to day and over the seasons. Weather can be described by measurable quantities, such as temperature, wind direction and speed, and precipitation.*

Science
Earth science
severe weather

Integrated Processes
Observing
Identifying
Comparing and contrasting

Materials
Game cards

Background Information
Severe weather can take many forms including hurricanes, tornadoes, thunderstorms, and flash floods. Each of these have unique dangers, but in every case there are things you can do to be prepared and stay as safe as possible when in severe weather conditions. Students should be aware of the steps they can take and the things they can encourage their families to do to minimize damage and injury during severe weather.

Management
1. Each group of two to four students will need one set of playing cards. The cards should be copied onto card stock and laminated, if possible.

Procedure
1. Ask students what kinds of severe weather events they can think of. Record their responses on the board. Be sure that the list includes hurricanes, tornadoes, thunderstorms, and flash floods.
2. Discuss the differences between these kinds of severe weather, and have students identify what they know about the various kinds of storms.
3. Have students share what they know about the safety precautions they should take during these kinds of situations. Record their responses on the board. Be sure to discuss all of the items listed on the game cards.
4. Explain that students will be playing a game to help them think about and remember some severe weather safety and preparedness tips.
5. Have students get into their groups and distribute one set of playing cards to each group. Instruct students to shuffle the cards and lay them face down in a grid.
6. Explain that one person will begin by turning over any two cards. If those cards show a safety precaution and its description, the player gets to remove the cards and turn over two more. If the cards do not match, it is the next player's turn.
7. Tell them to continue playing until all of the cards have been matched up.

8. When groups have completed their games, tell them that they get one point for every pair of cards. The student with the most points wins.
9. Play as desired to reinforce the safety precautions for severe weather conditions.

Connecting Learning

1. What kinds of severe weather can you think of? [hurricanes, tornadoes, thunderstorms, flash floods]
2. What do you need to do to prepare for severe weather? [make a disaster kit, have a family emergency plan, etc.]
3. What things should you do to stay safe during a tornado? ...thunderstorm? ...flash flood? ...etc.
4. Why is it important to know what to do during a storm?
5. What are you wondering now?

Internet Connections

FEMA for Kids
http://www.fema.gov/kids/
Information for kids on how to be prepared for disasters and prevent disaster damage.

NOAA Office of Climate, Water, and Weather Services
http://www.nws.noaa.gov/om/severeweather/index.shtml
Severe weather awareness and preparedness information.

Centers for Disease Control and Prevention
http://www.bt.cdc.gov/disasters/
Emergency preparedness and response information for natural disasters and severe weather.

American Red Cross
http://www.redcross.org/
Information on preparing at home, at school, at work, and in the community for emergencies and severe weather.

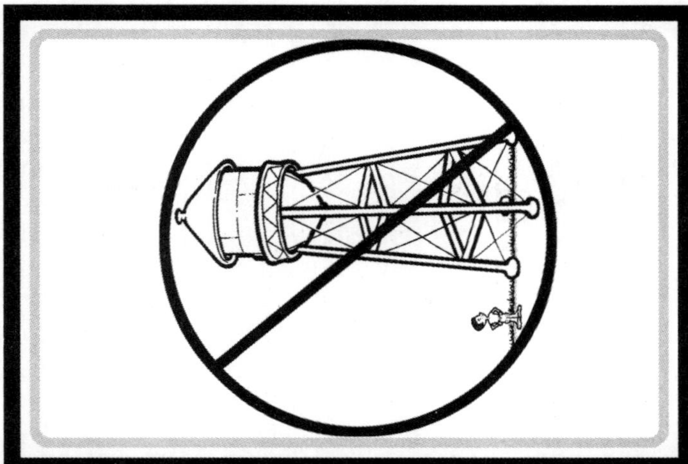

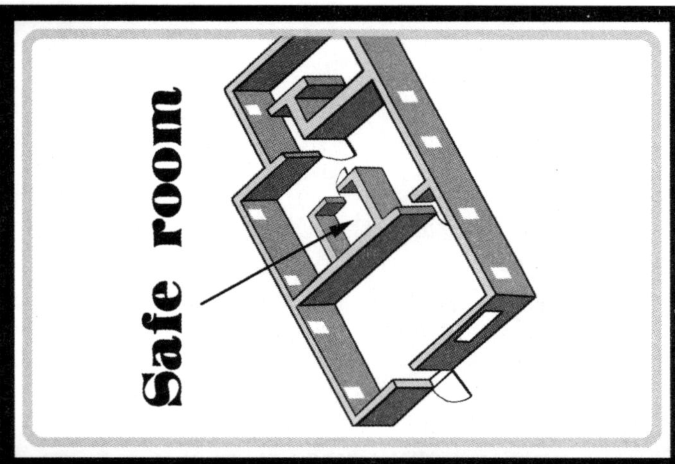
Safe room

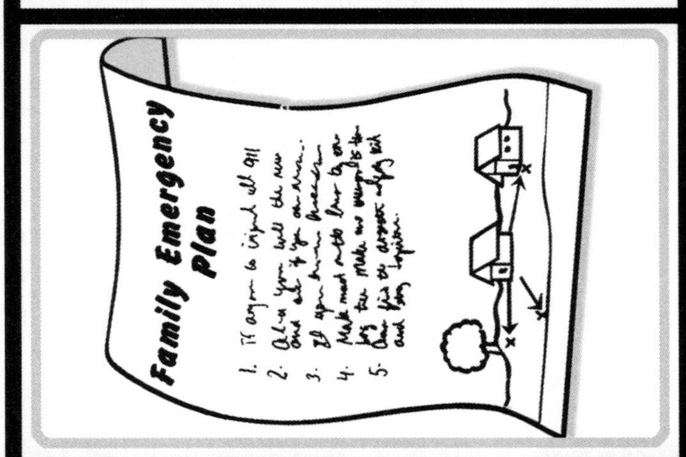
Family Emergency Plan

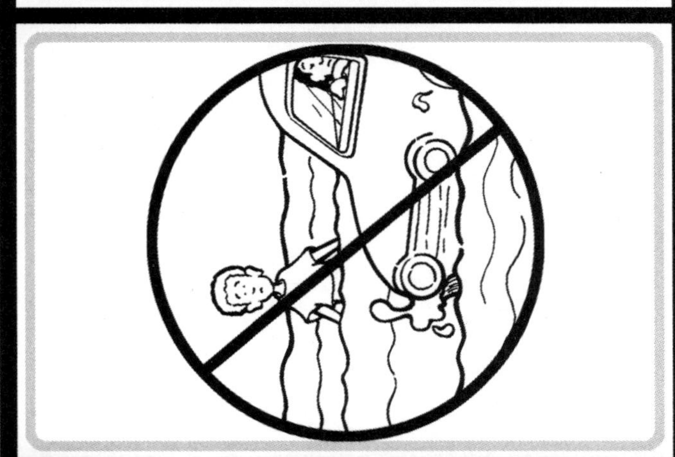

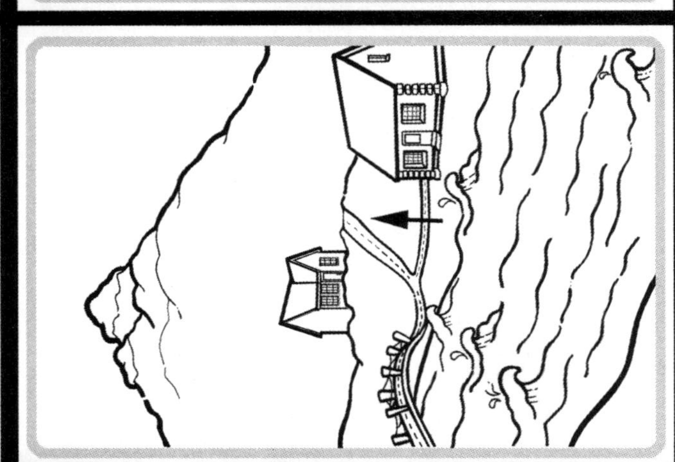

Thunderstorms

Stay away from tall objects outside.

Thunderstorms

Do not use the telephone.

Thunderstorms

Unplug appliances.

Thunderstorms

Close windows.

Tornadoes

Go to the basement or a safe room.

Tornadoes

Do not stay in a car or mobile home.

Flash Floods

Do not walk, swim, or drive in flood waters.

Flash Floods

Move to higher ground.

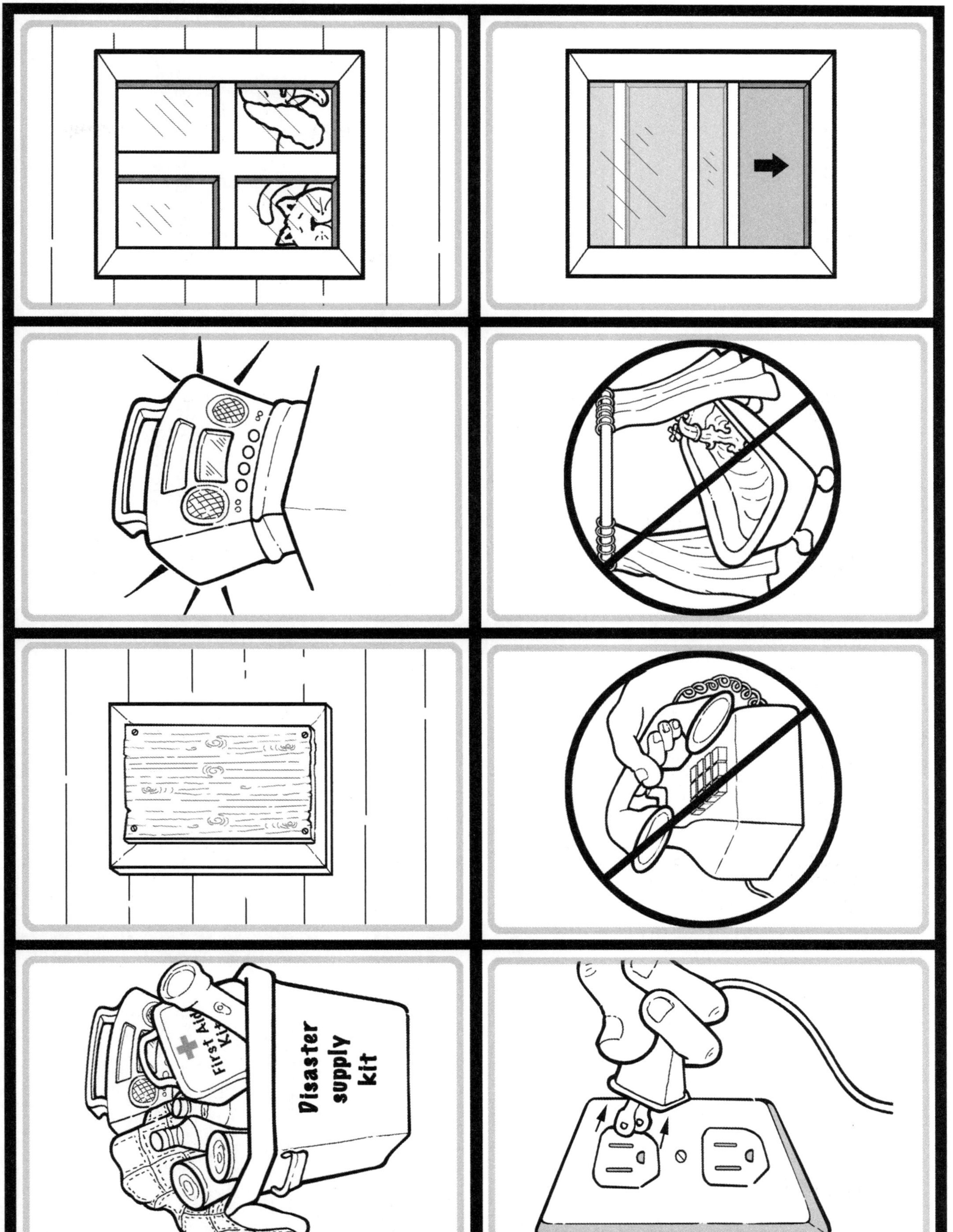
First Aid Kit
Disaster supply kit

Prepare

Make a disaster supply kit.

Prepare

Have a family emergency plan.

Prepare

Listen to the weather radio.

Hurricanes

Board up windows.

Tornadoes, Hurricanes, Thunderstorms

Bring your pets inside.

Tornadoes, Hurricanes, Thunderstorms

Stay away from windows.

Thunderstorms

Go inside if you see lightning or hear thunder.

Thunderstorms

Do not take a bath or shower.

Weather Words

Topic
Weather

Key Question
How can we act out weather vocabulary to help us better understand what the words mean?

Learning Goal
Students will recall weather terminology including temperature, wind speed, wind direction, precipitation, rain, snow, sleet, and hail.

Guiding Documents
Project 2061 Benchmarks

- *Tools such as thermometers, magnifiers, rulers, or balances often give more information about things than can be obtained by just observing things without their help.*
- *Describing things as accurately as possible is important in science because it enables people to compare their observations with those of others.*
- *Some events in nature have a repeating pattern. The weather changes some from day to day, but things such as temperature and rain (or snow) tend to be high, low, or medium in the same months every year.*

NRC Standard

- *Weather changes from day to day and over the seasons. Weather can be described by measurable quantities, such as temperature, wind direction and speed, and precipitation.*

Science
Earth science
  weather

Integrated Processes
Observing
Comparing and contrasting
Drawing conclusions
Generalizing

Materials
Colored paper (see *Management 1)*
Cotton balls
Aluminum foil
Glue
Scissors
Tape
String
Sentence strips
Student assessment page

Background information

It is important for young children to have a basic understanding of weather conditions and the vocabulary associated with those conditions. This activity will address the following terms: *wind speed, wind direction, temperature, precipitation, hail, snow, rain,* and *sleet.*

Wind direction refers to the direction from which the wind is coming, not where it is going. It is identified using wind vanes and reported as one of the eight directions of a compass (N, NE, E, SE, S, SW, W, NW). *Wind speed* is the strength or force of the wind. It is measured by an *anemometer* and, in the United States, is usually reported in miles per hour. *Temperature* tells how hot or cold something is. The warmer something feels, the higher the temperature is likely to be. A *thermometer* is an instrument whose size, shape, or some other feature changes when its temperature changes so that it can be used to measure temperature. *Rain, snow, sleet,* and *hail* are all forms of *precipitation. Precipitation* is simply any form of water that falls to the Earth's surface.

Management

1. The class will need scraps of bulletin board paper and/or construction paper and cardboard to make weather props such as raindrops, clouds, snowflakes, etc.
2. This activity should follow activities in which students have collected weather data.

Procedure

1. Ask the *Key Question* and state the *Learning Goal.*
2. Discuss and define the following weather words based on previous experiences: snow, sleet, rain, hail, temperature, wind direction, wind speed, and precipitation. Place these terms on sentence strips and place the words and definitions in a pocket chart.
3. Allow the students to break into groups of three. Assign each group a weather word. Tell the students that they will be responsible for explaining and illustrating their assigned weather word using props and actions. Provide time for each group to make props that will help others understand the weather terms. Non-fiction weather books should be available for inspiration.
4. When props are finished and each group has a good understanding of its term, provide time for the groups to share.
5. Lead a discussion about what the students learned from the experience.
6. Distribute the student assessment page. Ask students to identify the correct term represented in each picture. Have them give the meaning in their own words on the lines provided.

Connecting Learning

1. What is precipitation?
2. How did acting out weather vocabulary words help you understand their meanings?
3. How could you show other students what wind direction is and how it is measured?
4. What are you wondering now?

Extensions

1. Give students newspapers and have them cut out weather words and pictures to create a weather collage or weather picture dictionary.
2. Create a play using the weather props.

Weather Words

temperature, wind speed, precipitation, rain, snow, sleet, hail, and wind direction

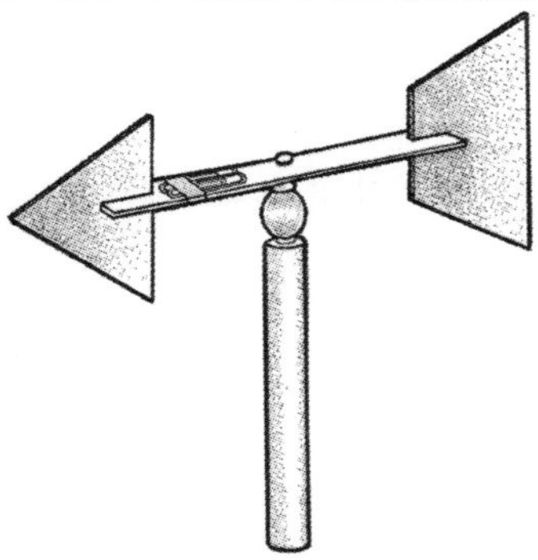

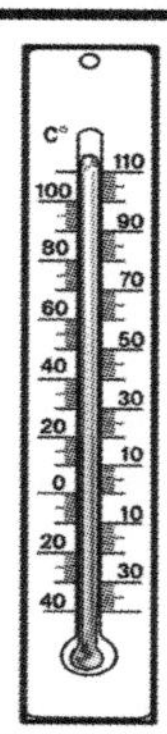

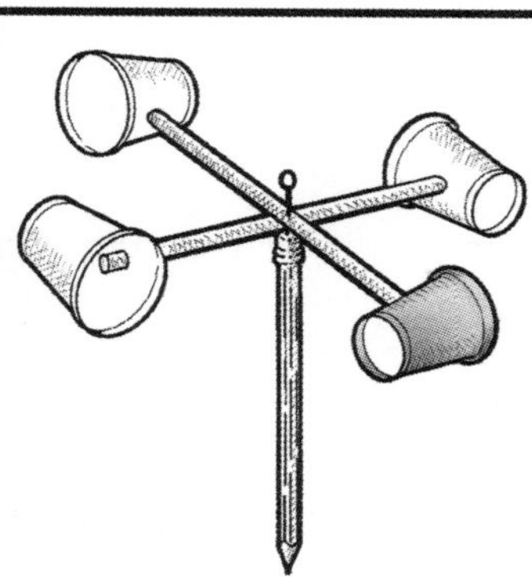

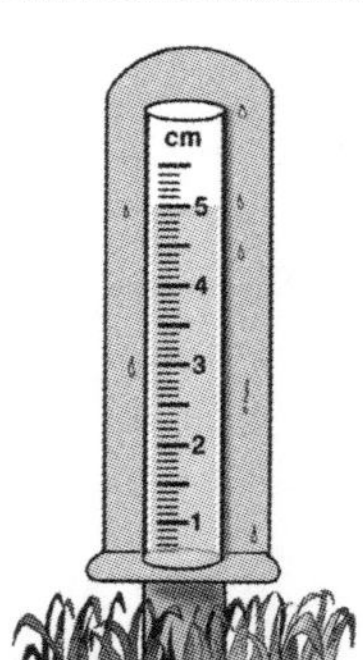

Weather Words

temperature, wind speed, precipitation, rain, snow, sleet, hail, and wind direction

Weather Bibliography

Bauer, Caroline Feller. *Windy Day.* J.B. Lippincott. New York. 1988.

Borden, Louise W. *Caps, Hats, Socks and Mittens.* Scholastic, Inc. New York. 1992.

Branley, Franklyn M. *Down Comes the Rain.* HarperTrophy. New York. 1997.

Branley, Franklyn M. *Flash, Crash, Rumble, and Roll.* HarperTrophy. New York. 1999.

Branley, Franklyn M. *Tornado Alert.* HarperTrophy. New York. 1990.

Burke, Jennifer S. *Hot Days.* Children's Press. Chicago. 2000.

Burke, Jennifer S. *Cloudy Days.* Children's Press. Chicago. 2000.

Burke, Jennifer S. *Cold Days.* Children's Press. Chicago. 2000.

Burke, Jennifer S. *Rainy Days.* Children's Press. Chicago. 2000.

Burke, Jennifer S. *Sunny Days.* Children's Press. Chicago. 2000.

Burke, Jennifer S. *Windy Days.* Children's Press. Chicago. 2000.

Carle, Eric. *Little Cloud.* Putnam Juvenile. New York. 2001.

Dickstein, Leslie. *Time for Kids: Storms.* HarperTrophy. New York. 2006.

Dorros, Arthur. *Feel the Wind.* HarperTrophy. New York. 1990.

Eckart, Edana. *Watching the Weather.* Children's Press. Chicago. 2004.

Eckart, Edana. *Watching the Seasons.* Children's Press. Chicago. 2004.

Fowler, Allan. *How Do You Know It's Winter?* Children's Press. Chicago. 1991.

Fowler, Allan. *How Do You Know It's Fall?* Children's Press. Chicago. 1992.

Fowler, Allan. *How Do You Know It's Spring?* Children's Press. Chicago. 1991.

Fowler, Allan. *How Do You Know It's Summer?* Children's Press. Chicago. 1992.

Fowler, Allan. *What' the Weather Today?* Children's Press. Chicago. 1991.

Fowler, Allan. *When a Storm Comes Up.* Children's Press. Chicago. 1995.

Gibbons, Gail. *The Seasons of Arnold's Apple Tree.* Voyager Books. New York. 1988.

Gibbons, Gail. *Weather Words and What They Mean.* Holiday House. New York. 1992.

Hall, Marie Ets. *Gilberto and the Wind.* Penguin Puffin Books. New York. 1978.

Hutchins, Pat. *The Wind Blew.* Aladdin. New York. 1993.

Llewellyn, Claire. *Wild, Wet, and Windy.* Scholastic, Inc. New York. 1998.

Rabe, Tish. *Oh Say Can You Say What's the Weather Today?: All About Weather.* Random House. New York. 2004.

Root, Phyllis. *One Windy Wednesday.* Candlewick Press. Cambridge, MA. 1997.

Shaw, Charles G. *It Looked Like Spilt Milk.* HarperCollins. New York. 1947.

Trumbauer, Lisa. *What Is a Thermometer?* Children's Press. Chicago. 2004.

White, Nancy. *The Magic School Bus Kicks Up a Storm.* Scholastic, Inc. New York. 2000.

Young, June. *Look How It Changes!* Children's Press. Chicago. 2006.

The AIMS Program

AIMS is the acronym for "**A**ctivities **I**ntegrating **M**athematics and **S**cience." Such integration enriches learning and makes it meaningful and holistic. AIMS began as a project of Fresno Pacific University to integrate the study of mathematics and science in grades K-9, but has since expanded to include language arts, social studies, and other disciplines.

AIMS is a continuing program of the non-profit AIMS Education Foundation. It had its inception in a National Science Foundation funded program whose purpose was to explore the effectiveness of integrating mathematics and science. The project directors, in cooperation with 80 elementary classroom teachers, devoted two years to a thorough field-testing of the results and implications of integration.

The approach met with such positive results that the decision was made to launch a program to create instructional materials incorporating this concept. Despite the fact that thoughtful educators have long recommended an integrative approach, very little appropriate material was available in 1981 when the project began. A series of writing projects ensued, and today the AIMS Education Foundation is committed to continuing the creation of new integrated activities on a permanent basis.

The AIMS program is funded through the sale of books, products, and professional-development workshops, and through proceeds from the Foundation's endowment. All net income from programs and products flows into a trust fund administered by the AIMS Education Foundation. Use of these funds is restricted to support of research, development, and publication of new materials. Writers donate all their rights to the Foundation to support its ongoing program. No royalties are paid to the writers.

The rationale for integration lies in the fact that science, mathematics, language arts, social studies, etc., are integrally interwoven in the real world, from which it follows that they should be similarly treated in the classroom where students are being prepared to live in that world. Teachers who use the AIMS program give enthusiastic endorsement to the effectiveness of this approach.

Science encompasses the art of questioning, investigating, hypothesizing, discovering, and communicating. Mathematics is a language that provides clarity, objectivity, and understanding. The language arts provide us with powerful tools of communication. Many of the major contemporary societal issues stem from advancements in science and must be studied in the context of the social sciences. Therefore, it is timely that all of us take seriously a more holistic method of educating our students. This goal motivates all who are associated with the AIMS Program. We invite you to join us in this effort.

Meaningful integration of knowledge is a major recommendation coming from the nation's professional science and mathematics associations. The American Association for the Advancement of Science in *Science for All Americans* strongly recommends the integration of mathematics, science, and technology. The National Council of Teachers of Mathematics places strong emphasis on applications of mathematics found in science investigations. AIMS is fully aligned with these recommendations.

Extensive field testing of AIMS investigations confirms these beneficial results:

1. Mathematics becomes more meaningful, hence more useful, when it is applied to situations that interest students.
2. The extent to which science is studied and understood is increased when mathematics and science are integrated.
3. There is improved quality of learning and retention, supporting the thesis that learning which is meaningful and relevant is more effective.
4. Motivation and involvement are increased dramatically as students investigate real-world situations and participate actively in the process.

We invite you to become part of this classroom teacher movement by using an integrated approach to learning and sharing any suggestions you may have. The AIMS Program welcomes you!

Get the Most From Your Hands-on Teaching

When you host an AIMS workshop for elementary and middle school educators, you will know your teachers are receiving effective, usable training they can apply in their classrooms immediately.

AIMS Workshops are Designed for Teachers
- Hands-on activities
- Correlated to your state standards
- Address key topic areas, including math content, science content, and process skills
- Provide practice of activity-based teaching
- Address classroom management issues and higher-order thinking skills
- Include $50 of materials for each participant
- Offer optional college (graduate-level) credits

AIMS Workshops Fit District/Administrative Needs
- Flexible scheduling and grade-span options
- Customized workshops meet specific schedule, topic, state standards, and grade-span needs
- Sustained staff development can be scheduled throughout the school year
- Eligible for funding under the Title I and Title II sections of No Child Left Behind
- Affordable professional development—consecutive-day workshops offer considerable savings

Call us to explore an AIMS workshop
1.888.733.2467

Online and Correspondence Courses

AIMS offers online and correspondence courses on many of our books through a partnership with Fresno Pacific University.
- Study at your own pace and schedule
- Earn graduate-level college credits

See all that AIMS has to offer—visit us online

http://www.aimsedu.org

Check out our website where you can:
- preview and purchase AIMS books and individual activities;
- learn about State-Specific Science and Essential Math;
- explore professional development workshops and online learning opportunities;
- buy manipulatives and other classroom resources; and
- download free resources including articles, puzzles, and sample AIMS activities.

find us on facebook

Become a fan of AIMS!
- Be the first to hear of new products and programs.
- Get links to videos on using specific AIMS lessons.
- Join the conversation—share how you and your students are using AIMS.

AIMS Education Foundation

Hands-On Math and Science

AIMS FOR YOU

SIGN UP TODAY!

While visiting the AIMS website, sign up for our FREE *AIMS for You* e-mail newsletter to get free activities, puzzles, and subscriber-only specials delivered to your inbox monthly.

AIMS Program Publications

Actions With Fractions, 4-9
The Amazing Circle, 4-9
Awesome Addition and Super Subtraction, 2-3
Bats Incredible! 2-4
Brick Layers II, 4-9
The Budding Botanist, 3-6
Chemistry Matters, 5-7
Counting on Coins, K-2
Cycles of Knowing and Growing, 1-3
Crazy About Cotton, 3-7
Critters, 2-5
Earth Book, 6-9
Electrical Connections, 4-9
Energy Explorations: Sound, Light, and Heat, 3-5
Exploring Environments, K-6
Fabulous Fractions, 3-5
Fall Into Math and Science*, K-1
Field Detectives, 3-6
Floaters and Sinkers, 5-9
From Head to Toe, 5-9
Getting Into Geometry, K-1
Glide Into Winter With Math and Science*, K-1
Gravity Rules! 5-12
Hardhatting in a Geo-World, 3-5
Historical Connections in Mathematics, Vol. I, 5-9
Historical Connections in Mathematics, Vol. II, 5-9
Historical Connections in Mathematics, Vol. III, 5-9
It's About Time, K-2
It Must Be A Bird, Pre-K-2
Jaw Breakers and Heart Thumpers, 3-5
Looking at Geometry, 6-9
Looking at Lines, 6-9
Machine Shop, 5-9
Magnificent Microworld Adventures, 6-9
Marvelous Multiplication and Dazzling Division, 4-5
Math + Science, A Solution, 5-9
Mathematicians are People, Too
Mathematicians are People, Too, Vol. II
Mostly Magnets, 3-6
Movie Math Mania, 6-9
Multiplication the Algebra Way, 6-8
Out of This World, 4-8
Paper Square Geometry:
The Mathematics of Origami, 5-12
Popping With Power, 3-5
Positive vs. Negative, 6-9
Primarily Bears*, K-6
Primarily Magnets, K-2
Primarily Physics: Investigations in Sound, Light, and Heat Energy, K-2
Primarily Plants, K-3
Primarily Weather, K-3
Problem Solving: Just for the Fun of It! 4-9
Problem Solving: Just for the Fun of It! Book Two, 4-9
Proportional Reasoning, 6-9
Puzzle Play, 4-8
Ray's Reflections, 4-8
Sensational Springtime, K-2
Sense-able Science, K-1
Shapes, Solids, and More: Concepts in Geometry, 2-3
The Sky's the Limit, 5-9
Soap Films and Bubbles, 4-9
Solve It! K-1: Problem-Solving Strategies, K-1
Solve It! 2nd: Problem-Solving Strategies, 2
Solve It! 3rd: Problem-Solving Strategies, 3
Solve It! 4th: Problem-Solving Strategies, 4
Solve It! 5th: Problem-Solving Strategies, 5
Solving Equations: A Conceptual Approach, 6-9
Spatial Visualization, 4-9
Spills and Ripples, 5-12
Spring Into Math and Science*, K-1
Statistics and Probability, 6-9
Through the Eyes of the Explorers, 5-9
Under Construction, K-2
Water, Precious Water, 4-6
Weather Sense: Temperature, Air Pressure, and Wind, 4-5
Weather Sense: Moisture, 4-5
What's Next, Volume 1, 4-12
What's Next, Volume 2, 4-12
What's Next, Volume 3, 4-12
Winter Wonders, K-2

Essential Math

Area Formulas for Parallelograms, Triangles, and Trapezoids, 6-8
Circumference and Area of Circles, 5-7
Effects of Changing Lengths, 6-8
Measurement of Prisms, Pyramids, Cylinders, and Cones, 6-8
Measurement of Rectangular Solids, 5-7
Perimeter and Area of Rectangles, 4-6
The Pythagorean Relationship, 6-8

* Spanish supplements are available for these books. They are only available as downloads from the AIMS website. The supplements contain only the student pages in Spanish; you will need the English version of the book for the teacher's text.

For further information, contact:
AIMS Education Foundation • P.O. Box 8120 • Fresno, California 93747-8120
www.aimsedu.org • 559.255.6396 (fax) • 888.733.2467 (toll free)

Duplication Rights

No part of any AIMS books, magazines, activities, or content—digital or otherwise—may be reproduced or transmitted in any form or by any means except as noted below.

Standard Duplication Rights

- A person or school purchasing AIMS activities (in books, magazines, or in digital form) is hereby granted permission to make up to 200 copies of any portion of those activities, provided these copies will be used for educational purposes and only at one school site.
- For a workshop or conference session, presenters may make one copy of any portion of a purchased activity for each participant, with a limit of five activities or up to one-third of a book, whichever is less.
- All copies must bear the AIMS Education Foundation copyright information.
- Modifications to AIMS pages (e.g., separating page elements for use on an interactive white board) are permitted only within the classroom or school for which they were purchased, or by presenters at conferences or workshops. Interactive white board files may not be uploaded to any third-party website or otherwise distributed. AIMS artwork and content may not be used on non-AIMS materials.

Standard duplication rights apply to activities received at workshops, free sample activities provided by AIMS, and activities received by conference participants.

Unlimited Duplication Rights

Unlimited duplication rights may be purchased in cases where AIMS users wish to:
- make more than 200 copies of a book/magazine/activity,
- use a book/magazine/activity at more than one school site, or
- make an activity available on the Internet (see below).

These rights permit unlimited duplication of purchased books, magazines, and/or activities (including revisions) for use at a given school site.

Activities received at workshops are eligible for upgrade from standard to unlimited duplication rights.

Free sample activities and activities received as a conference participant are not eligible for upgrade from standard to unlimited duplication rights.

State-Specific Science modules are licensed to one classroom/one teacher and are therefore not eligible for upgrade from standard to unlimited duplication rights.

Upgrade Fees

The fees for upgrading from standard to unlimited duplication rights are as follows.
For individual activities, the cost is $5 per activity per school site.
For Literature Links bundles, the cost is $12 per bundle per school site.
For books, the cost is based on the price of the book (see table).

Book Price	Upgrade Fee
$9.95	$15.00/site
$18.95	$24.00/site
$21.95	$27.00/site
$24.95	$30.00/site
$29.95	$35.00/site
$34.95	$40.00/site
$49.95	$55.00/site

The cost of upgrading is shown in the following examples:
For five activities at six schools:
5 activities x $5 x 6 schools = $150

For two books (at $21.95) at 10 schools:
2 books x $27 x 10 schools = $540

For three books (at $24.95) and four activities at eight schools:
(3 books x $30 x 8 schools) + (4 activites x $5 x 8 schools) = $720 + $160 = $880

Purchasing Unlimited Duplication Rights

To purchase unlimited duplication rights, please provide us the following:
1. The name of the individual responsible for coordinating the purchase of duplication rights.
2. The title of each book, activity, and/or magazine issue to be covered.
3. The number of school sites and name and address of each site for which rights are being purchased.
4. Payment (check, purchase order, credit card).

Requested duplication rights are automatically authorized with payment. The individual responsible for coordinating the purchase of duplication rights will be sent a certificate verifying the purchase.

Internet Use

AIMS materials may be made available on the Internet if all of the following stipulations are met:
1. The materials to be put online are purchased as PDF files from AIMS (i.e., no scanned copies).
2. Unlimited duplication rights are purchased for all materials to be put online for each school at which they will be used. (See above.)
3. The materials are made available via a secure, password-protected system that can only be accessed by employees at schools for which duplication rights have been purchased.

AIMS materials may not be made available on any publicly accessible Internet site.